SENSOR FUSION

Volume 931

Contents

(continued)

Proceedings of SPIE—The International Society for Optical Engineering

Volume 931

Sensor Fusion

Books are to be returned on or before
the last date below.

Fi   la
                                                                                                                iiversity

LIBREX—

Published by

SPIE—The International Society for Optical Engineering
P.O. Box 10, Bellingham, Washington 98227-0010 USA
Telephone 206/676-3290 (Pacific Time) • Telex 46-7053

SPIE (The Society of Photo-Optical Instrumentation Engineers) is a nonprofit society dedicated to advancing engineering
and scientific applications of optical, electro-optical, and optoelectronic instrumentation, systems, and technology.

The papers appearing in this book comprise the proceedings of the meeting mentioned on the cover and title page. They reflect the authors' opinions and are published as presented and without change, in the interests of timely dissemination. Their inclusion in this publication does not necessarily constitute endorsement by the editors or by SPIE.

Please use the following format to cite material from this book:
   Author(s), "Title of Paper," *Sensor Fusion,* Charles B. Weaver, Editor, Proc. SPIE 931, page numbers (1988).

Library of Congress Catalog Card No. 88-61020
ISBN 0-89252-966-0

Printed in the United States of America.

SENSOR FUSION

Volume 931

Conference Committee

Chair
Charles B. Weaver
Honeywell Electro-Optics Division

Cochairs
Elizabeth Beggs
MAC Associates, Inc.

Harley E. Manley
Grumman Aerospace Corporation

Session Chairs
Session 1—Sensor Fusion Concepts
Harley E. Manley, Grumman Aerospace Corporation

Session 1 (continued)—Sensor Fusion Concepts
Elizabeth Beggs, MAC Associates, Inc.

Session 2—System Aspects
Charles B. Weaver, Honeywell Electro-Optics Division

Session 2 (continued)—System Aspects
Alan N. Steinberg, The Analytic Sciences Corporation

Session 3—Analysis
Charles B. Weaver, Honeywell Electro-Optics Division

Conference 931, *Sensor Fusion,* was part of a five-conference program on Infrared Imaging and Thermal Sensing held at SPIE's 1988 Technical Symposium on Optics, Electro-Optics, and Sensors. The other conferences were

Conference 929, *Infrared Optical Materials VI*
Conference 930, *Infrared Detectors and Arrays: Critical Reviews of Optical Science and Technology*
Conference 933, *Multispectral Image Processing and Enhancement*
Conference 934, *Thermosense X.*

Program Chair: **George Zissis,** Environmental Research Institute of Michigan

SENSOR FUSION

Volume 931

INTRODUCTION

Sensor fusion is becoming increasingly important as limitations in performance of single function sensors become apparent. The limitation of real estate for sensors from medical operating rooms to military platforms is also driving the requirement for sensors to share a combination of assets such as housings, optics, detectors, signal processors, displays, and controls. As a consequence, integration of several technologies into a sensor suite has proven to be important in several fields.

Advanced aircraft are using central processors to process information from a number of different sensors to improve performance, reliability, maintainability, and survivability. Optical systems are required to work at several different spectral bands and to coexist with sensors of other technologies in such diverse areas as submarine periscopes to space platforms. Progress in multifunction technology has been rapid and forms a very interesting area for discussion.

This proceedings consists of 30 papers containing the most recent research on sensor fusion, and thus serves as a valuable contribution to the development of the technology. My thanks to the cochairs, session chairs, and authors for bringing their knowledge and skills to this forum.

Charles B. Weaver
Honeywell Electro-Optics Division

SENSOR FUSION

Volume 931

Session 1

Sensor Fusion Concepts

Chair
Harley E. Manley
Grumman Aerospace Corporation

"Fusion – the key to tactical mission success"

Vito G. Comparato Director, Advanced Systems,

Fairchild Weston Systems, Inc.
300 Robbins Lane, Syosset, New York 11791

ABSTRACT

Next generation tactical missions cannot rely on threat information or countermeasure actions which depend solely on one sector of the electromagnetic spectrum. Rather, they must rely on an integrated, multi-sensor approach for "engagement information" extraction.

Current programs such as Pave Pillar identify fusion as the key to the required positive and unambiguous situation and target assessment. A look at the fusion problem is presented which includes a conceptual description & functional flow.

1.0 INTRODUCTION

Current individual (stand-alone) sensor systems do not meet the projected operational requirements for real-time, long-range tactical scenarios. The major mission requirements have been identified as system survivability, low observables, interleaved acquisition, track and engagement modes, target classification and identification. All presume high engagement support rates. First generation system concepts, built around individual sensor capabilities, have failed to meet the performance requirements outlined above for the expected mission environments. The problems have apparently stemmed, not from individual sensor deficiencies, but more so from the philosophy taken in utilizing individual sensors.

This paper discusses the issues and elements of multisensor fusion from the perspective of what is likely to be accomplished on-board our next generation tactical platforms.

Representative platforms include:

Air Force – Advanced Tactical Fighter (ATF), Navy – Advanced Tactical Aircraft (ATA), Army –Armored Family of Vehicles (AFV), and any of a variety of Remotely Piloted Vehicles/Unmanned Aeriel Vehicles (RPV/UAV).

Although the discussions will draw heavily from the world of Airborne Tactical missions and have, perhaps, an EW perspective, we believe that the concepts are applicable to the non-airborne tactical environment as well, and actually span the issues of total mission functions including surveillance, threat warning and fire control.

2.0 OUR CUSTOMERS' COMPOUND PROBLEM

Ever-escalating mission objectives and associated subsystem costs no longer permit our tactical missions the luxury of procuring a large variety of independent "Black Boxes" which independently address requirements. In the face of fixed force structure and numerical disadvantages, integrated systems are required in order to field smaller, less costly suites. In fact programs such as Pave Pillar attest to the fact that our customers are rethinking their organization and procurement options in order to field systems in which surveillance, threat warning, fire control and countermeasures are tightly coupled together.

Multisensor fusion is not just a buzz word or desirement. Rather it is a firm requirement – prototype equipment is currently under contract.

The joint services have recognized the potential performance improvements available by utilizing complementary sensors in an integrated manner. Specifically, multisensor data correlation would aid in target classification, track maintenance, and graceful system degradation. Conceptually, the multisensor solution appears promising, since it implies a much higher density of "observables", i.e., information about targets of interest. In addition to the increased quantity of data which multisensor systems provide, the diversity in the spectral, as well as temporal, characteristics of the different sensors may be further exploited to yield not only useful target discriminants for classification, but near immunity to countermeasures, such as ECCM and EOCM, which significantly degrade single-sensor system performance.

The realities, then, of the threat environment, the mission requirement, the low observables doctrine, and the acquisition philosophy drive us to multisensor fusion. The resultant expectation is one of a greater probability of mission success and survivability while attaining to a higher degree than ever before the key capabilities of high availability, reliability, maintainability, testability, and future growth.

3.0 MULTISENSOR SYSTEM

Figure 1 represents a generalized multisensor system. The system includes a variety of RF, IR, and EO sensors for threat warning (TW), search and track (ST), and fire control. The system concept is to perform as many functions as possible at the sensor level. This accrues many advantages such as minimum time-latency and easier reconfiguration for fault tolerance. The preprocessors are shown as distinct nodes in order to allow for "specialization" and uniqueness that specific sensors might require. The fusion processor combines the reports/tracks from the different sensors (including off-board remoted data via data link) to enhance the target (threat) classification and state estimation. The fusion functions result in highly accurate and confident identification in order to properly select responses to the assessed situation and allocate resources such as sensors and countermeasures.

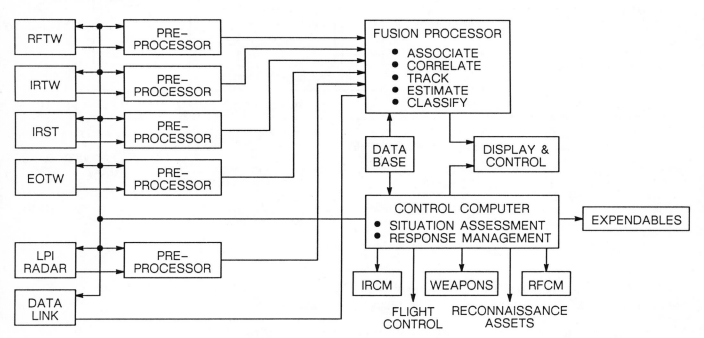

Figure 1. Multi-sensor system functions

Measurement data is provided by each sensor independently. Tracking is performed autonomously when gains in target state estimation or threat identification can be realized. The trackers are tailored to the sensor. Sensor data from non-EW sensors and from off-board platforms is included and is available as track files and can be used by fusion after alignment. The combination of EW, non–EW, and off–board track files and reports provides the data stream to the fusion function.

The system and the fusion function operate in one of two modes depending on the threat environment: surveillance or threat warning. In the first mode (surveillance), the system is in a low stress situation, probably in passive operation. The fusion function is receiving data from all onboard and off–board sensors and is performing the functions of coordinate alignment, time propagation, association, correlation and track update. Classification of the threat information is attempted on a sensor basis where possible and is refined in fusion using all of the available sensor data. Some sensors can provide good classification autonomously while other sensors provide accurate angle or range data but no classification. The fusion of the data will combine the attributes of each sensor in one master file. The updated master track file (including classification data) is then transmitted to the situation assessment function, the integrated display subsystem and the fire control subsystem. In the second mode (threat warning), the system is under stress due to the identification of an impending threat. The two modes are non–exclusive, i.e., the surveillance mode operates continuously and the threat warning operates on a demand basis.

3.1 The fusion function

Figure 2 decomposes fusion into its related subfunctions. Each sensor is assumed to work autonomously and asynchronously in time and space. Furthermore, corresponding to each sensor are detection and track level preprocessors.

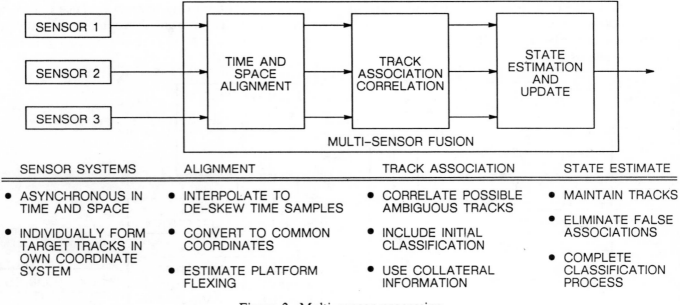

Figure 2. Multi–sensor processing

Individually formed target tracks are provided by each sensor to the fusion function in each sensor's own coordinate system. Fusion then performs the following six subfunctions:

- time propagation – track files which have state vectors and error covariance matrices will be "time propagated" to the fusion update time.

- coordinate alignment – the sensor data (including track files) need to be referenced to a common origin and compensation provided to sensor misalignment.

- association – the tracks and reports from different sensor files need to be compared to determine candidates for "fusion"; attribute tests, kinematic tests, and probabilistic tests are utilized.

- <u>correlation</u> – the results of the association are then processed to determine the track pairs which will be fused.

- <u>track update</u> – the tracks which "best correlate" are used to update the corresponding state vector and error covariance matrix; the output is an updated composite (master) track file.

- <u>classification</u> – the tracks are examined and an assessment is made in an attempt to determine the target type, lethality, and threat priority.

The fusion function extends from the measurement data to the master track file and until threat warning indications have been developed. For each sensor, signal processing is necessary. The next step is track processing, which is adapted to each sensor in order to produce either track file or a time history of reports. Time propagation and coordinate alignment create data with common reference points in time and space which is necessary to blend the data. This blending or fusion of the data is performed in order to create one master track file with all the positive attributes of each sensor. This master track file is then available to situation assessment, the pilot's display, and for an optional function of transfer to the fire control system. Threats identified as "high priority" can be processed immediately and transmitted to the situation assessment function. Fusion also performs sensor control, monitoring, and cueing. Alternative algorithms may be invoked by decisions within the adaptive control function based on both the system and sensor states.

3.2 Overall requirements are satisfied

By examining the goals of our multisensor system against the fusion capabilities, the merits of the approach will be evident.

The ultimate goal of our tactical multisensor system is to accomplish its mission with an increased probability of survival, $P(s)$. Fusion should have a major impact on these values. In the approach presented here, the sensors can operate in both the autonomous and integrated modes. This architecture is key to the fault tolerance aspects, but also serves to enhance the classification and location problem. By using the capability of each sensor by itself to perform classification and tracking of targets and identifying lethal threats at the sensor level, the system can be very responsive. Then by combining the results of all sensors, both surveillance and threat warning, a better "understanding" of the environment can be developed within the fusion function prior to passing information to situation assessment. Although not explicitly a part of fusion, the optimization of $P(s)$ will be derived from the accurate threat classification and location developed in the fusion function.

Fusion itself can operate in an "autonomous" mode, utilizing data from on-board sensors only. However, the system can also perform in a "full-up" configuration integrating both the on-board and off-board information. This approach can offer leverage with the multisensor system utilizing this data by "fusing" the reports and tracks with on-board sensor data.

3.3 Fusion is fault tolerant

The technical approach presented in this report is inherently fault tolerant for the following reasons:

- Tailored track filters for each sensor can operate autonomously as required, and perform classification and identification.

- The master track file is created from all available data, i.e., if one sensor fails, the track file simply does not use that data and if the sensor returns, the data can again be used without reconfiguration. In essence, the fusion function will be adaptable to whatever sensors are available.

- Fusion will also be adaptable to the treat environment. In a very dense environment and in threat warning situations, the time consuming association and correlation processing can be bypassed. Fusion, then, performs the time propagation and common reference alignment and then passes data on to the situation assessment for immediate attention and response. Thus, fusion if properly applied will not fail from data overload and will, in fact, reduce the data rates sufficiently for the situation assessment and external functions.

Figure 3 illustrates one aspect of the fault tolerance which will be built into fusion. Multiple paths exist for the data to propagate from the sensors to the situation assessment and other functions.

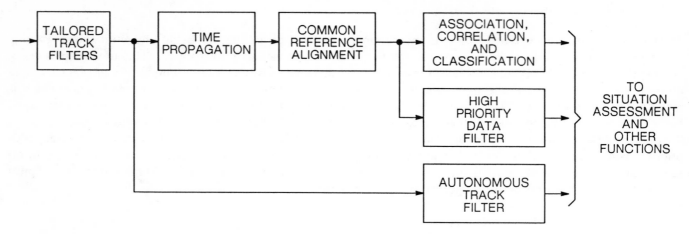

Figure 3. Fusion is fault tolerant

3.4 Fusion interface

The fusion function, and more generally the whole multisensor system will, in the 1990's time-frame, integrate into the host platform such that the interfaces are well specified but functions probably will be shared. For example, in terms of fusion, multiple levels of fusion will undoubtedly be performed in the host platform to blend all different subsystem sensors. The radar and other offensive sensors will be combined with the defensive sensors, possibly for fire control and display functions. Thus, given all the different subsystems and sensors which will be installed on future tactical platforms, the fusion of data will be a more general function than one exclusive part of any subsystem.

Given the above understanding, the fusion function can be separately identified and interfaces to it can be defined, as shown in Figure 4. ICNIA provides the off-board data (e.g., via JTIDS), navigation data, and aircraft identification (IFF) data. The radar, and infrared search and track (IRST) will be interfaced directly. The EW sensors include all sensors forming a part of the EW suite of equipment.

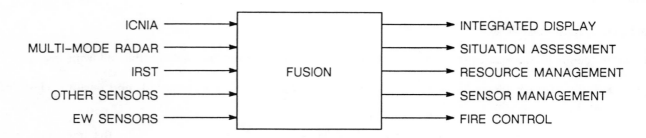

Figure 4. Fusion interface

The integrated display will receive updated reports of threat and emitter locations and other warning and status data. The situation assessment function will receive a list of all active emitters via the master track file. A tentative priority by lethality will have been established by the fusion function (including those threats with ambiguous lethality) but the situation assessment function will provide final priority rankings and initiate responses. Resource management will maintain the status of all EW resources and thus will receive from fusion the status of fusion functions as well as sensor performance. Sensor control refers to sensor cueing and pointing. The fire control interface is optional but would consist of accurate threat location, etc., from the master track file.

3.5 Fusion is adaptive

In order to handle the future threat environments, fusion will need to be adaptable to an ever changing threat environment. Fusion must be responsive to quickly changing battle conditions. For example, given the current capability of the red forces to use their electromagnetic, infrared and optical search and track techniques from the same weapon system requires that the fusion be responsive to all modes of the hostile system. In addition, the multisensor system must be able to perform the threat classification and location on many threats simultaneously, each threat using any of the various search and track sensors. Fusion will be optimized by use of a control function to adapt the system to, for example, different threats by using different filter models or algorithms.

The Fusion Adaptive Control function, illustrated in Figure 5, will control the time updates and algorithm selection for the fusion function as well as provide sensor status and sensor cueing commands. The need for this type of function is clear.

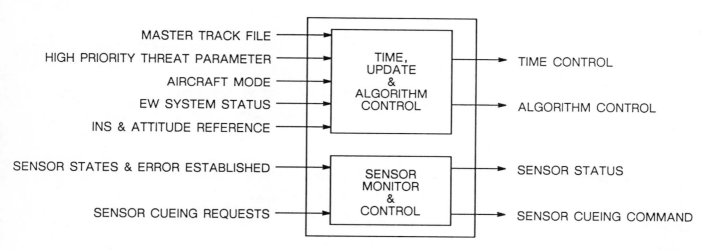

Figure 5. Fusion is adaptive to the demands of the environment

4.0 CONCLUSIONS

With the approaching availability of VLSI and VHSIC processing capabilities and architectures, the ability to physically integrate two or more advanced sensor classes on a common platform will become a reality. The significant result of these developments is the capability to perform real-time multisensor data processing through post-detection and beyond, without sacrificing the individual sensor "quasi-autonomous" status. Furthermore, this allows mission-driven selection and combination of these complementary sensor detection-level parameters to assure the "optimal" instantaneous performance, requirements for low observables, and countermeasure immunity are satisfied.

Multisensor information fusion for target detection and classification

Michael C. Roggemann, James P. Mills, Steven K. Rogers, Matthew Kabrisky

Air Force Institute of Technology, Departments of Engineering Physics
and Electrical Engineering, Wright-Patterson AFB OH 45433

ABSTRACT

Merging information available from multisensor views of a scene is a useful approach to
target detection and classification. Development of multisensor information fusion techniques using a data base of real imagery from an absolute range laser radar and a corresponding forward looking infrared (FLIR) sensor is underway. Our conceptual approach to
multisensor target detection and classification uses sensor-dependent segmentation and
feature extraction. Information is fused first at the detection level and then within the
classifier. We hypothesize that an approach to information fusion based on the mathematical theory of evidence (i.e., evidential reasoning) is a useful method for multisensor
object classification. In this paper we summarize an approach to a multisensor object
classification system, discuss results of multisensor segmentation algorithm, and present
an evidential reasoning-based approach to a multisensor classifier.

1. INTRODUCTION

Fusing the outputs of multiple sensors offers a promising approach to automatically
detecting and estimating the class membership of objects in a scene. We are studying
theoretical issues and algorithimic approaches to detecting and classifying objects of
interest in natural scenes using absolute range images and intensity images. A data base
of real images from a colocated absolute range laser radar sensor and a forward-looking
infrared (FLIR) sensor images is available to support this research.

This paper begins with a discussion of the approach and system level considerations for
the multisensor object recognition system being developed. Segmentation techniques and
object detection approaches are presented next. Features and classifier considerations are
then discussed.

2. APPROACH

A conceptual approach to a multisensor object recognition system is shown in Figure 1.
The system is considered to have n sensors. Sensor-dependent segmentation and feature
extraction are provided for each sensor. Segmentation information is provided to an
executive control routine. The executive negotiates the regions passed by segmentation
across the sensors to declare tentative detections, providing the first level of information fusion. Sensor-dependent feature extraction is performed on regions approved by the
executive. The classifier accepts feature values from the extraction subsystems and
commands from the executive, and determines an estimate of the class membership of candidate regions in the scene.

Processing begins with sensor dependent operations. Each sensor has an associated
segmentation system. Ideally, the output of the segmentation is a collection of regions
which correspond exactly to regions in the image which contain targets. In fact, the
segmentation process often passes many collections of pixels which do not correspond to an
object of interest. Many of these pixels may be excluded from further processing by
simple tests. Some of these tests can be performed as sensor-dependent processing while
others require information fusion.

Sensor-dependent segmentation reports are passed to the executive control routine,
henceforth referred to as the executive. The first role of the executive is to negotiate
segmentation reports from the sensor-dependent subsystems and arrive at a concensus of
detected regions. Geometrical parameters for all sensors, such as pointing information and
resolution, are provided to the executive. The executive first seeks to register segmentation reports in a geometrical space. Segmentation reports are automatically labeled as
detections when spatial correspondences across the sensors are found. Conflicts arise
when not all segmentation reports passed by individual sensors can be registered with
segmentation reports from other sensors. A simple approach to resolving conflicts is to
make the assumption that segmented regions which cannot be registered across the sensors
are uninteresting, and discard these regions from further consideration. One potentially
more robust approach, which is more computationally expensive, is to relax the segmentation
criteria in regions corresponding to segmentation reports in the other sensor-dependent

subsystems. New segmented regions would typically result from this activity, which would need to be subjected to the process outlined above. Another approach to resolving conflicts is to evaluate the confidence that a region passed by some, but not all, sensor dependent segmentation subsystems is a real object of interest.

The executive orders feature extraction on regions passing the detection tests. Feature values are passed to the multisensor classifier for interpretation.

3. PREPROCESSING AND SEGMENTATION

In this section, techniques found useful for preprocessing and segmenting the infrared and range images in the database are summarized.

Isolated noise pixels are present in both the infrared imagery and the range imagery. Standard median filtering and a local averaging filter, using 3x3 pixel regions in both cases, have provided sufficient smoothing. Images processed in this fashion are called filtered images.

Filtered infrared images are segmented using a technique first proposed by Hamadani[1], which is summarized here. The Hamadani algorithm is an adaptive approach to segmenting regions in an intensity image which are brighter than the background. The pixel mean and standard deviation of the filtered image are computed. A new image is also computed from the filtered image by applying a modification of the Kirsch edge operator[2] to the filtered image, referred to as the Kirsch image. The Kirsch operator, applied to all 3x3 pixel regions in an image, involves assigning a value in a new image in the position of the center pixel of each 3x3 region, P, the value given by:

$$P = \max(1, \max_{i=0}^{i=7} k_1 \times (a_{i+1} + a_{i+2}) - k_2 \times (a_{i+3} + a_{i+4} + a_{i+5} + a_{i+6} + a_{i+7})) \tag{1}$$

where the a_i's are assigned as in Fig. 2 and all subscripts are evaluated modulo eight. Kirsch's original work used $k_1 = 5$ and $k_2 = 3$. Hamadani found $k_1 = 10$ and $k_2 = 1$ more suitable for his database. The pixel mean and standard deviation of the Kirsch image are also computed.

Thresholds for both the filtered infrared image and the Kirsch image are computed according to the rules:

$$\text{Thresh}_f = \text{mean}_f + k_f \times \text{sd}_f$$

$$\text{Thresh}_k = \text{mean}_k + k_k \times \text{sd}_k \tag{2}$$

where the subscripts refer to either the filtered or Kirsch image, sd represents the standard deviation, and k_f and k_k are constants. Values of $k_f = 2.3$ and $k_k = 1.9$ are satisfactory for the present database. Pixels in the filtered and Kirsch images are compared to the thresholds according to the rule applied to corresponding pixels in the filtered and Kirsch images:

```
    if (filtered image pixel > thresh_f) and
       (Kirsch image pixel   > thresh_k) then
         pixel is passed for further processing
    else
         pixel is excluded from further processing
    and if
```
(3)

Small, isolated collections of pixels which pass this test are then rejected.

Good results have been achieved using this approach on the infrared imagery available. Fig. 3 shows representative cases of raw and segmented FLIR images. Objects brighter than the background are reliably found. However, other objects or regions brighter than the background will also be passed by this approach. More sophisticated means are required to reject these unwanted regions.

Range image segmentation is accomplished by fitting planes to the three dimensional surface implied by an absolute range image. Information in range images is stored in the form azimuth angle (az), elevation angle (el), range, where az and el are inferred from knowledge of the angular sampling rate of the sensor and the (row, column) position of the pixel in the image. Taking the center of the image as az = el = 0 (corresponding to the boresight of the sensor), a straightforward geometric transformation converts (az, el, range) maps into Cartesian three-dimensional (x,y,z) maps of the reflectors in the scene. Since the objects to be found are manmade metal objects (i.e., vehicles) viewed in natural scenes, a segmenter which seeks out the planar quality of the candidate objects is used. The plane-fitting approach has been suggested by others, including Mitiche and Aggarwal[3] for the purpose of finding edges in range images. In the present case, planar surfaces rather the edges are sought. The technique is summarized here.

The segmentation algorithm begins by computing an (x,y,z) map of the (az, el, range) image provided. Planes are fit, in the least-squares sense, to 3x3 regions of the (x,y,z) map and the error associated with the plane fit is computed. An array whose elements are given by the error associated with the least-squares plane fit at the center of the 3x3 region, called the error image, is searched for regions where high accuracy planes were fit (i.e., where the error of the plane fit was low). The pixels in the error image are thresholded according to the rule:

```
if (error pixel < thresh) then
   pixel is passed for further processing
else
   pixel is excluded from further processing
end if                                                                    (4)
```

The threshold can be chosen to be very small if truely planar objects (i.e., planar on the scale of the spatial sampling of the sensor) are expected: for the present case a threshold equivalent to a total absolute error (i.e., the absolute value of the error of the least squares plane fit summed over all nine pixels in the 3x3 region) of 1 m was used.

The pixels in the filtered image corresponding to the pixels in the error image which passed the above test are placed in a new array, called temp. Isolated small groupings of pixels are rejected from temp. The regions remaining in temp are grown by one pixel. Growing by one pixel compensates for the large error associated with the case when a 3x3 region used for plane fitting was only partially contained in the object, causing the outermost pixels of these regions to be lost through the thresholding operation. The regions in temp are then examined individually. The mean range of each region is computed to scale a size window. Regions which pass the test:

(min height < region height < max height) and
(min width < region width < max width)   (5)

where the height and width are computed in pixels, are considered to be potential objects of interest. Max height and width are the largest height and width of any object of interest as determined from the absolute dimensions of the objects. Min height and width are determined similarly.

Good results have been achieved using this approach on the database. Regions containing objects of interest are extracted quite reliably. Fig. 4 shows a typical instance of a filtered absolute range image and the results of applying the segmentation algorithm. In Fig. 4 the pixels in the filtered image have been remapped using the rule display pixel = remainder (filtered range pixel/256) to fit the entire dynamic range of the image on an eight bit display. Occasionally, regions not containing objects of interest are also passed. These regions must be examined with more robust techniques.

The next processing step is to register information about the regions passed by the infrared and range image segmenters to negotiate the reports of regions of interest and declare object detections. In many instances the non-object regions passed by the sensor-dependent systems are not from corresponding areas in the actual scene. Exploiting this observation in software will demonstrate the first level of information fusion and should yield a powerful approach to object detection. Current research is focused on this aspect of the approach.

4. FEATURES AND CLASSIFIER CONSIDERATIONS

Features based on intensity and two and three dimensional shape are being developed for use by the classifier. Examples of features include hot spot evaluation, moments, size, and relative location and orientation of planes composing objects.

The classifier algorithm being studied is based on the mathematical theory of evidence, or evidential reasoning, as developed by Shafer[4]. Multisensor decision automation using evidential reasoning approaches has been suggested by others including Lowrance and Garvey[5], and Bogler[6]. This theory is well suited to the present case, where the a priori statistics (i.e., the probability distribution functions of the various feature values, and the functional dependence of these distribution functions) are unknown, and perhaps unknowable. We now summarize an approach to an evidential reasoning-based multisensor classifier.

In place of the more conventional approach based on approximating probability distribution functions, evidential reasoning (in the context of current problem) treats feature measurements as pieces of evidence. Feature measurements are initially evaluated against a space of possibilities called the frame of discernment. The frame of discernment is taken to be a set of n+1 propositions of the form "the region being evaluated contains object i", 1<=i<=n, where n is the number of objects of interest, plus the proposition "the region being evaluated does not contain an object of interest". The features are evaluated through functions called either "basic probability assignments"[4] or "mass distributions"[5]. The "mass distribution" terminology is adopted here for brevity.

Mass distributions assign a portion of total belief to subsets of the frame discernment through functional dependence on the feature values. Separate mass distributions must be developed for interpreting the output of each sensor-dependent feature extractor. The functional form of the mass distributions is application dependent, and subjective knowledge may be incorporated.

Once determined, the mass distributions are combined, or pooled, into a single mass distribution through Dempster's rule of combination (DRC). DRC is a technique for combining arbitrarily complex mass distributions into a single mass distribution which represents a concensus of the individual mass distributions[4]. This new mass distribution is used to interpret evidence and compute the evidential interval [spt(A), pls(A)] for subsets, A, of the frame of discernment. Spt(A) is the support for A, defined as the degree to which the evidence supports the proposition "A is true". Pls(A) is the plausibility of A, defined as the degree to which the evidence fails to refute the proposition "A is true". Complete ignorance is represented by the evidential interval [0,1]. Precise likelihood assignments, such as would result from a conventional decision approach, would be represented spt(A)=pls(A). Degrees of ignorance (i.e., lack of conclusive evidence) are represented by evidential intervals between these two extremes[4].

Decisions are made by evaluating the evidential intervals resulting from applying DRC. Criteria for rendering the ultimate decision are application dependent. Future research will explore mass functions and choice of decision criteria.

5. ACKNOWLEDGEMENTS

The data used to support this research was gathered and provided by the Army Night Vision and Electro-Optics Center (NVEOC), Ft. Bellvior VA. The authors extend their thanks to two divisions of NVEOC, the Advanced Concepts Division and the Laser Division for the time and effort spent assisting this project.

6. REFERENCES

1. Hamadani, Naser A. "Automatic Target Cueing in IR Imagery", Masters Thesis, Air Force Institute of Technology, Wright-Patterson AFB, Ohio 1981.
2. Kirsch, Russell A., "Computer Determination of the Constituent Structure of Biological Images", Computers and Biomedical Research, Vol. 4, 1971, pp. 315-328.
3. Mitiche, Amar, and J. K. Aggarwal, "Detection of Edges Using Range Information, IEEE Transactions on Pattern Analysis and Machine Intelligence, PAMI-5, No. 2, March 1983, pp. 174-178.
4. Shafer, Glenn, A Mathematical Theory of Evidence, Princeton University Press, Princeton NJ, 1976.
5. Lowrance, John D., and Thomas D. Garvey, "Evidential Reasoning: An Implementation for Multisensor Integration", SRI International Menlo Park CA, 1983.
6. Bogler, Philip L., "Shafer-Dempster Reasoning with Applications to Multisensor Target Identification Systems", IEEE Transactions on Systems, Man, and Cybernetics, Vol. SMC-17, No. 6, November/December 1987, pp. 968-977.

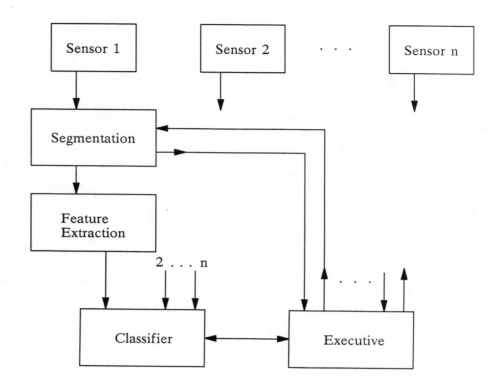

Figure 1. Multisensor system block diagram.

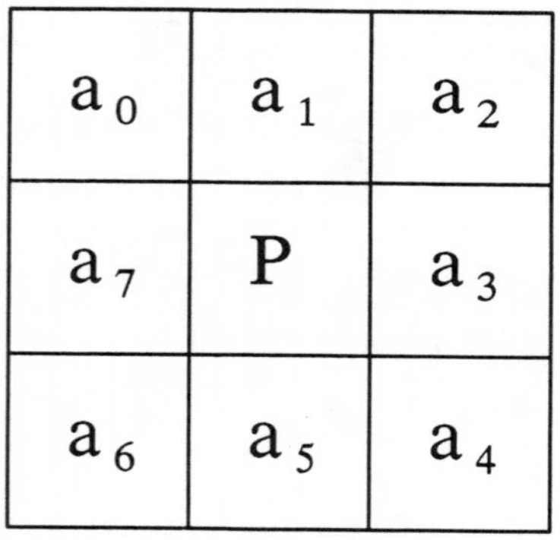

a_0	a_1	a_2
a_7	P	a_3
a_6	a_5	a_4

Figure 2. Assignment of pixels for Kirsch operators.

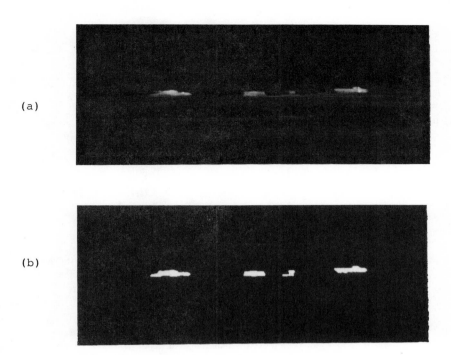

Figure 3.   Representative example of raw FLIR image (a) and segmentation output (b).

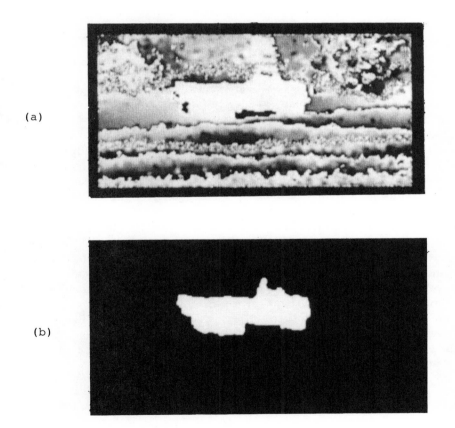

Figure 4.   Representative example of raw absolute range image with modulo 256 performed
            for display (a) and segmentation output (b).

Multisensor Target Detection and Classification

Dennis W. Ruck   Steven K. Rogers   Matthew Kabrisky

James P. Mills

Air Force Institute of Technology

AFIT/ENG, Wright-Patterson AFB, Ohio 45433

Abstract

In this paper a new approach to the detection and classification of tactical targets using a multifunction laser radar sensor is developed. Targets of interest are tanks, jeeps, trucks, and other vehicles. Doppler images are segmented by developing a new technique which compensates for spurious doppler returns. Relative range images are segmented using an approach based on range gradients. The resultant shapes in the segmented images are then classified using Zernike moment invariants as shape descriptors. Two classification decision rules are implemented: a classical statistical nearest-neighbor approach and a multilayer perceptron architecture.

The doppler segmentation algorithm was applied to a set of 180 real sensor images. An accurate segmentation was obtained for 89 percent of the images. The new doppler segmentation proved to be a robust method, and the moment invariants were effective in discriminating the tactical targets. Tanks were classified correctly 86 percent of the time. The most important result of this research is the demonstration of the use of a new information processing architecture for image processing applications.

1 Introduction

The problem of machine interpretation of images remains an unsolved problem. Past work has met with modest success in limited cases. Previous attempts at machine vision have utilized only a single sensor. The weaknesses of the sensor have resulted in machines with less robust vision than desired. Hence, current work is emphasizing the integration of information from multiple sensors. The strengths of one sensor will be used to offset the weaknesses of the other sensors resulting in robust machine vision.

The goal of this research was to extend current work on multisensor fusion at the Air Force Institute of Technology by providing algorithms for detection and classification of targets using a multifunction laser radar device. This paper will describe the new algorithms developed for the detection and classification of tactical targets using doppler and relative range images. The results presented were obtained by applying the algorithms to a set of actual sensor images. The tactical targets of interest in this research were M60 tanks, 1.25-ton trucks, 2.5-ton trucks, jeeps, and Petroleum, Oil, and Lubricants tanker trucks (POLs).

The detection and classification of targets is basically a three step process. In the first step, a source image is segmented to find possible targets. The segmentation of the doppler images was performed using three new methods while the relative range images were segmented using a previously developed range gradient-based technique [8] [9] [5]. The second step is extraction of shape features (a feature vector) for each distinct region in the segmented image. Zernike moment invariants were used for shape descriptors. Finally, the feature vectors are classified. Two classification methods were used: a traditional statistical decision rule and an artificial neural network.

The next section will review background material necessary to understanding the algorithms developed in section 3. The results of applying these algorithms to sensor data will be presented in section 4, and conclusions will follow in section 5.

2 Background

This section will cover three topics. First, the data used for developing and testing the images will be described. Then the use of moments, in general, and Zernike moments, in particular, will be reviewed for shape analysis. Finally, the application of multilayer perceptrons to the classification of feature vectors will be detailed.

The data used for the development and testing of the algorithms was collected from targets consisting of M60 tanks, 1.25-ton and 2.5-ton trucks, jeeps, and Petroleum, Oil, and Lubricants tankers (POLs). The multifunction laser radar provided registered pixel data for doppler, relative range, passive visible, passive infrared, and carrier intensity images. Each image consisted of 256 x 256 pixels with eight bits per pixel. For this research effort, only the doppler, relative range, and carrier intensity images were utilized.

Moments have previously been used for target classification [1]. An ordinary moment is defined below

$$M_{pq} = \int_{-\infty}^{\infty} \int_{-\infty}^{\infty} x^p y^q f(x,y) dx dy \tag{1}$$

When the function $f(x,y)$ is binary, the moment gives information about the shape of the object represented by $f(x,y)$ [4].

The moments defined by equation 1 will change in value as the object changes in position, rotation, and scale. It is desirable to have the features be invariant under these changes. The variance under position is easily eliminated by computing the centralized moments. The centralized moments replace the kernel $x^p y^q$ by $(x - \bar{x})^p (y - \bar{y})^q$ where $(\bar{x}, \bar{y})$ is the centroid of the object. The variance under changes in scale is eliminated by dividing the moment by a factor which is proportional to the size of the object. For example, scale normalized moments can be defined as shown below

$$\mu'_{pq} = \frac{\mu_{pq}}{\mu_{00}^{(p+q+2)/2}} \tag{2}$$

where μ'_{pq} is the scale normalized moment and μ_{pq} is the centralized moment [7]. The variance due to rotation is more difficult to eliminate. Dudani describes in [1] how combinations of the scale normalized moments can be made invariant under rotation. The term moment invariant is used for these combinations of scale normalized moments which are invariant under changes in position, scale, and rotation.

The Zernike moment invariants used in this research are similar to the moments already described with the exception that the kernel function $x^p y^q$ is replaced by a Zernike polynomial. They are also invariant under changes in position, scale, and rotation [7]. The Zernike moment invariants were computed from the scale normalized central moments shown in equation 2 using the equations given by Teague [7].

A multilayer perceptron was one of the methods used for classifying the feature vectors. A multilayer perceptron is a network of simple computing elements (often called nodes or, inappropriately, neurons). The nodes are connected in a feed forward fashion as shown in Figure 1.

The multilayer perceptron implements a decision rule by partitioning the vector space occupied by the feature vectors. The input to the network is a feature vector which has been appropriately normalized, and the output of the network indicates the classification of the feature vector. The ideal output is for a single node in the output to be active and all others inactive.

Each node in the network is a simple summing junction where each of the inputs is weighted according to a strength of connection. The weighted sum plus a threshold term is input to a sigmoid function. Initially, the weights and thresholds in the network are set to random values. A training procedure is then used to update the weights and thresholds. In this research, the Back Propagation algorithm originally developed by Werbos and rediscovered by Parker [3] was implemented. Paul J. Werbos developed this algorithm for his thesis, *Beyond Regression: New Tools for Prediction and Analysis in the Behavioral Sciences*, while a PhD student at Harvard University in 1974 [3]. The Back Propagation algorithm uses the difference between the desired output vector and the actual output vector as a cost function and iterates on a set of training vectors until the weights and thresholds in the network stabilize.

3 Algorithms

This section will describe the algorithms developed to perform target detection and classification on doppler and relative range images. The overall algorithm consists of three steps. The first step performs a segmentation of the source image. The second step takes the segmented image, locates the distinct regions, and calculates a set of features (a feature vector) for each region based on its shape. Finally, the feature vectors are evaluated to determine from which class of objects the feature vector originated. Each of these steps will now be briefly described.

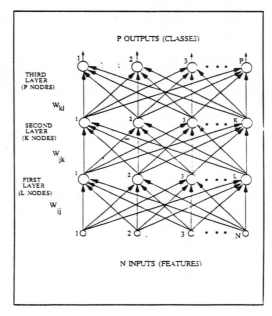

Figure 1: Multilayer Perceptron

3.1 Segmentation

The doppler images were segmented using algorithms developed during this research project while the relative range images were segmented using an algorithm developed by Tong [8] [9] [5]. The doppler segmentation algorithms will be described followed by a brief summary of the Tong algorithm.

Three techniques for segmenting the doppler images were developed and tested with sensor data of vehicles. The techniques will be described in the order in which they were developed.

The Optimum Thresholding (OT) Technique was developed using decision theory as a basis (see [2] and [6] for more information). It assumes that the doppler image is essentially composed of two separate distributions and that the distributions are Gaussian and of equal variance. Under these assumptions, the optimum threshold is the mean of the two modes in the histogram of the image. To find these modes, the technique first computed the average of the image from the histogram. Then the histogram was split into two parts about the average. The mode in each of the parts of the histogram was taken to be the desired values. This technique worked well except when the assumption of a bimodal distribution was not satisfied. This situation occurred primarily when the laser beam was specularly reflected from surfaces in the field of view of the sensor. The resultant loss in received carrier signal strength caused spurious doppler values which nullified the validity of the bimodal distribution assumption; hence, more robust methods were sought.

The Optimum Thresholding Using Carrier Intensity (OTUCI) technique was next developed to compensate when the carrier dropped out due to specular reflection. This method is identical to the previous method with the exception that during processing if the carrier intensity for a pixel is below a preset threshold, then that pixel is ignored. The joint histograms between received carrier intensity and doppler return value were examined for a group of images exhibiting carrier dropout. It was determined that a fixed threshold of 130 would be sufficient for the majority of cases. This algorithm proved to be relatively immune to carrier dropout in terms of the number of background pixels labeled target pixels; however, it failed to find the targets in a large number of test images; hence, another method was developed.

The Background Lobe Elimination (BLE) technique was developed to overcome the problems of carrier dropout while retaining sensitivity to the presence of targets. This method finds the lobe in the histogram due to the non-moving background and eliminates it. Carrier intensity is still used to determine whether or not a pixel is valid. After eliminating the background lobe, the algorithm searches the modified histogram for the new mode. The threshold for segmentation is then set to the average of the mode in the modified histogram (target mode) and the mode in the original histogram (background mode). This method proved to be highly robust even in the presence of severe noise due to carrier dropout while maintaining sensitivity to very small targets.

The relative range images were segmented using the technique developed by Tong [8]. The algorithm is based on the histogram of range gradients in an image. Targets are assumed to be relatively flat with concomitant low range gradients while the background is assumed to have much higher range gradients. An initial segmentation is performed based on an estimated threshold derived from the range gradient histogram. Then a relaxation process is applied to the segmented image to change the label assigned to a pixel when the neighboring pixels do not support that label. The algorithm has been shown to be highly robust in finding man-made targets using relative range imagery.

After the images are segmented, the next step in the algorithm is to detect the distinct regions and compute the shape features for each region.

3.2   Feature Extraction

The feature extraction process consists of three steps: 1) identify all distinct regions in a segmented image, 2) eliminate regions which are too small for shape analysis, and 3) compute the desired shape features for each remaining region. Each step will be described below.

The first step processes each segmented image and identifies all the distinct regions. The image is scanned in raster fashion until a target pixel is found. The border for that object is traversed, adding the border pixels to a list of pixels for that region. As each pixel is added to the border list, an auxiliary map is updated indicating the pixel has been added to a border list. After the border is traversed, the region is added to a list of regions for the image and scanning of the image restarts where it stopped until a new border pixel is found which has not been added to a list of pixels. When the entire image has been scanned, a list of regions containing the border pixels for each distinct region has been formed.

After creating the list of regions, the next step is to eliminate all regions which are too small for shape analysis. This process simply counts the number of border pixels for each region and eliminates those regions with fewer pixels than a preset threshold value. The threshold was set based on the number of pixels within the field of view and the minimum number of pixels on target required for identification.

The final step in the feature extraction process is to compute the shape features for each of the remaining regions. The Zernike moment invariants were computed from the central scale normalized moments using the formulas given by Teague [7]. Two different sets of moment invariants were computed. The first set was computed from the silhouette of the object, and the other from just the border pixels for the object as in [1]. Zernike moment invariants up to the fourth order were computed. Now that the feature vectors have been computed for the segmented regions, they must be compared to determine from which class they come. This process, known as classification, will be discussed next.

3.3   Classification

The process of classifying a feature vector was accomplished using two different methods. The first method was a traditional statistical classifier; and the other method used an artificial neural network, namely, a multilayer perceptron. Each method will be briefly described.

The statistical classification was accomplished in two parts: training and testing. For the training stage, two-thirds of the available feature vectors were used while reserving the remaining one-third for the testing stage. The training process consisted of three steps. First, the feature vectors were normalized to have zero mean and unity variance for each feature. This step assures that no single feature will dominate the decision due to scaling differences. The second step involved reducing the feature space dimensionality through the use of Fisher linear discriminants. Finally, the average feature vector was computed for each class.

Testing the statistical decision rule consisted of three steps. First, each vector was rescaled using the transformation computed in step one of the training process. Then each vector was mapped to a lower dimensionality space using the Fisher linear discriminants. Finally, the distance from the test vector to each class mean was computed, and the class with the lowest distance was selected as the class of the test vector.

The second method used for classifying the features was a multilayer perceptron. As with the nearest-neighbor classifier, there was a training stage and testing stage for the multilayer perceptron. The training stage consisted of two steps. The first step scales the training data so that all vectors lie within a unit hypercube where each component

Table 1: Doppler Segmentation Results - Target Localization

Quality	OT (%)	OTUCI (%)	BLE (%)
Excellent	84.4	0.0	71.1
Excellent-Good	0.6	0.0	7.2
Good	2.2	43.3	10.6
Good-Fair	0.0	1.7	3.3
Fair	0.0	1.1	3.9
Fair-Poor	0.6	0.6	0.6
Poor	12.2	53.4	3.4

of the training vectors lies between zero and one. This step guarantees that no single component dominates the classification process due to scaling factors and that the features are scaled similarly to the weights and thresholds in the network. The second step is to train the network using the Back Propagation algorithm. A training vector is selected at random from the pool of training vectors, and the error in the output of the multilayer perceptron is then used to update the weights. Training continues until the averaged distance between the output of the network and the desired output reaches a minimum.

Testing of the multilayer perceptron classifier is a three step process. First, the feature vector is mapped to another vector using the shifting and scaling determined during the training phase; then the shifted and scaled vector is presented to the network which computes the output. Finally, the output is classified using the following rule. If one of the outputs is above a preset within-class threshold and all other outputs are below a preset out-of-class threshold, then the output is declared unambiguous and the class corresponds to the node with the greatest output. If either the within-class threshold or the out-of-class threshold or both is not met, then the output is declared ambiguous and the class corresponds to the node with the largest output.

The results of applying these algorithms to the sensor data will be discussed next.

4   Segmentation and Classification Results

This section presents the results of applying the segmentation and classification algorithms to a set of real world data collected from a multisensor CO_2 laser radar. The segmentation results will be discussed first followed by the classification results.

4.1   Segmentation Results

The doppler segmentation results will be presented first followed by the relative range segmentation results. All the segmentations were judged by an observer who compared the segmented image with the source image. Each segmented image was given a rating indicating the proportion of target pixels correctly identified (target localization) and another rating for the number of incorrectly labeled background pixels (noise immunity).

The results of applying the three segmentation algorithms to 180 test images is given in Table 1 and 2. As shown in the tables, the Optimal Thresholding (OT) technique performed well in a majority of cases. Target localization was *good* or better for 87.2 percent of the test images, and noise immunity was *good* or better for 77.7 percent. The cases where the noise immunity was *poor* were characterized by moderate to severe carrier dropout; hence, the Optimal Thresholding Using Carrier Intensity (OTUCI) technique was developed.

The OTUCI technique was less immune to noise but was also less sensitive to the presence of targets. The noise immunity was *good* or better for 93.4 percent of the test images; but the target localization was only *good* or better for 43.3 percent - a significant reduction in target localization performance. The poor performance was due primarily to the use of a fixed threshold for all pixels. For some doppler images, there were spurious doppler returns with carrier intensities above the threshold. Since the spurious doppler returns are far from the zero motion return value, the spurious pixels have a large effect on the average computed for the image. Also, a large number of zero motion pixels are eliminated from the image due to carrier intensities below the threshold. By eliminating these zero motion

Table 2: Doppler Segmentation Results - Noise Immunity

Quality	OT (%)	OTUCI (%)	BLE (%)
Excellent	64.4	66.7	92.2
Excellent-Good	0.0	1.1	1.7
Good	13.3	25.6	4.4
Good-Fair	2.8	2.2	0.0
Fair	3.9	1.7	1.7
Fair-Poor	2.2	2.2	0.0
Poor	13.4	0.6	0.0

Table 3: Relative Range Segmentation - Target Localization and Noise Immunity

Quality	Target Localization (%)	Noise Immunity (%)
Excellent	0.0	6.8
Excellent-Good	1.4	6.8
Good	29.7	36.5
Good-Fair	13.5	12.2
Fair	33.8	23.0
Fair-Poor	4.0	9.4
Poor	17.6	5.4

pixels from the image average computation, the influence of the spurious pixels is magnified. Hence, another more robust method for doppler segmentation in the presence of carrier dropout was developed.

The Background Lobe Elimination (BLE) technique proved to be better than either of the previous techniques in both target localization and noise immunity performance. As shown in the tables, target localization was *good* or better for 88.9 percent of the test images which is slightly better than the Optimal Thresholding method and is much better than the Optimal Thresholding Using Carrier Intensity method. For noise immunity, the BLE technique was rated *good* or better for 98.3 percent of the images which is much better that the Optimal Thresholding technique and somewhat better than the OTUCI method.

The results of applying the relative range segmentation algorithm to 74 test images is given in Table 3. For use with a vehicle classifier, the relative range segmentation algorithm exhibited three main weaknesses: 1) splitting of targets, 2) noise from low range-gradient backgrounds, and 3) segmentation of buildings. The splitting of targets occurred when range ambiguity jumps were located on those targets. This splitting causes problems for the classifier which is based on the shape of single regions. The algorithm was originally developed to find all man-made objects, so it was not surprising that buildings were segmented as well. For this research, however, only mobile targets were of interest.

The segmented images were then processed by the feature extractor, and the resulting feature vectors were used to train and test the classifiers. The results of testing the classifiers will be discussed next.

4.2   Classification Results

Two classifiers were tested using the feature vectors derived from the segmented images. The statistical classification results will be discussed first followed by the multilayer perceptron results.

The statistical classifier was relatively accurate at classifying tanks; however, it did not perform as well with the other classes. Table 4 shows the confusion matrix for the nearest-neighbor classifier. Classification accuracy for tanks was 76.5 percent.

The multilayer perceptron classifier performed somewhat better than the nearest-neighbor classifier. The confusion matrix for this classifier is given in Table 5. For the multilayer perceptron the classification accuracy was 86.4

Table 4: Nearest-Neighbor Classifier Confusion Matrix. Reference Data (Columns) vs Test Data (Rows).

	Tank	Jeep	POL	Truck, 2.5 ton
Tank	13	3	1	0
Jeep	1	0	1	1
POL	3	0	0	2
Truck, 2.5 ton	0	0	1	1

Table 5: Multilayer Perceptron Classifier Confusion Matrix. Reference Data (Columns) vs Test Data (Rows).

	Unambiguous Output				Ambiguous Output	
	Tank	Jeep	POL	Truck, 2.5 ton	Correct	Incorrect
Tank	19	1	0	0	1	1
Jeep	2	0	0	0	0	1
POL	0	0	0	0	0	2
Truck, 2.5 ton	0	0	0	1	0	1
Clutter	10	2	0	0	N/A	9

percent. Since the output vector could be classified as *ambiguous*, a set of objects representing typical clutter from the relative range segmentations was input to the network during testing. The network correctly gave an ambiguous output for 43 percent of the test objects.

There are two likely reasons for the poor recognition capability for the classes other than *tank*. The first reason is that many of the silhouettes for the other classes had very poor shape due to the high range from which the images were taken. Many of the shapes were difficult even for human observers to correctly classify. The second reason is that the feature vector does not provide sufficient discrimination for the shapes. For example, the *POL* class was the front view of a POL and the *truck* class was also the front view. These classes were very similar in shape; hence, the discrimination would have to be based on fine detail which would require higher order moment invariants. Nonetheless, the classifier performed nearly as well as a human observer using only the segmented silhouettes for classification.

5 Conclusion

Two of the doppler segmentation algorithms developed were highly successful. The Background Lobe Elimination method was very robust. It was able to find targets even in the presence of severe carrier dropout noise. The nearest-neighbor and multilayer perceptron classifiers both performed well at classifying tanks. These results were obtained from actual sensor data which was collected in the field. The classifiers worked nearly as well as human observers given the same information. This research has shown that an artificial neural network is a viable classifier even in real world situations where the data are highly noisy and incomplete.

6 Acknowledgments

The authors wish to express their thanks to Mr. Dave Power of the Air Force Wright Aeronautical Laboratories (AFWAL/AARI) for his support of this research.

References

[1] Dudani, S.A. *et al.* "Aircraft Identification by Moment Invariants," *IEEE Transactions on Computers, C-26*:39-45 (January 1977).

[2] Gonzalez, Rafael C. and Paul Wintz. *Digital Image Processing.* London: Addison-Wesley Publishing Company, 1977.

[3] Parker, David B. "A Survey of Recent Work on Back Propagation," To be published in *Proceedings of IEEE Conference on Neural Information Processing Systems - Natural and Synthetic* (Novemeber 1987).

[4] Pavlidis, Theodosios. *Structural Pattern Recognition.* Berlin, Heidelberg, and New York: Springer-Verlag, 1977.

[5] Rogers, Steven K. *et al.* "Multisensor Fusion of Ladar and Passive Infra-Red Imagery for Target Segmentation." to be published in *Optical Engineering.*

[6] Ruck, Dennis W. *Multisensor Target Detection and Classification.* MS thesis, AFIT/GE/ENG/87D-56. School of Engineering, Air Force Institute of Technology (AU), Wright-Patterson AFB Ohio, December 1987.

[7] Teague, Michael Reed. "Image Analysis via the General Theory of Moments," *Journal of the Optical Society of America, 70*:920-930 (August 1980).

[8] Tong, Carl W. *Target Segmentation and Image Enhancement through Multisensor Data Fusion.* MS thesis, AFIT/GE/ENG/86D-55. School of Engineering, Air Force Institute of Technology (AU), Wright-Patterson AFB Ohio, December 1986.

[9] Tong, Carl W. *et al.* "Multisensor Data Fusion of Laser Radar and Forward Looking Infrared (FLIR) for Target Segmentation and Enhancement," *Infrared Sensors and Sensor Fusion*, Rudolph G. Busor, Frank B. Warren, Editors, Proc. SPIE 782, pp. 10-19 (1987).

A Geometric approach to multisensor fusion

Su-shing Chen

Department of Computer Science, University of North Carolina
Charlotte, NC 28223

Abstract

Multisensor fusion is by integrating several numerical and spatial sensory data to obtain more complete spatial information about the environment. The first step is to gather 3-D spatial information about the environment by partial surface reconstruction from each sensory image data. The second step is to integrate different 3-D spatial information derived from image data, considered as partial and incomplete information, by combination techniques.

1. Introduction

Multisensor fusion is by integrating several numerical (e.g. frequency or histogram) and spatial (e.g. 2-D and 3-D image) sensory data to obtain more complete and improved spatial information so that the environment can be understood. The first step is to gather 3-D spatial information about the environment by partial surface reconstruction from each sensory image data. The second step is to integrate different image data, considered as partial and incomplete information, by certain combination techniques.

We shall classify multisensor fusion into two categories: static and dynamic multisensor fusions. Static multisensor fusion is concerned with the first step of surface reconstruction of several sensory images of different type located at one point and the second step of integration on (the tangent planes) of a single image sphere.

Integration on the image sphere is also useful to temporal images of a moving target. This is the simpler kind of dynamic multisensor fusion. The other kind of dynamic multisensor fusion is investigated in [1] by active sensing in the region of focus of attention and by integration of multiframes of one or a group of objects in the region.

In this paper, we shall consider a multisensor family located at one viewing position. Such a multisensor family may be either a homogeneous

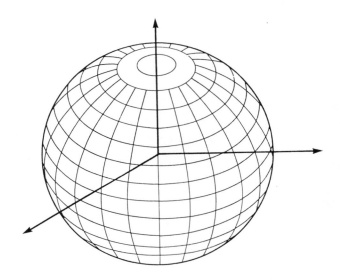

sensor array or a heterogeneous family of sensors. The spherical model [2],[3] eliminates several limitations of the usual orthographic and planar perspective models, including distortion and vanishing points.

The spherical model is independent of different sensor types. Although images of sonar, optical, radar, range and infrared sensings are of different natures, their reconstructed surfaces are of the same geometric representation in the 3-D world space. In this single model, some of the uncertainties and ambiguities due to modeling are eliminated.

2. Taxonomy of sensor fusion

Multisensor fusion is classified into two categories: (1) static multisensor fusion, and (2) dynamic multisensor fusion. In the first case, different sensory data about a static environment of possibly heterogeneous static sensors are integrated. Dynamic multisensor fusion has three possibilities: (a) objects are moving, but sensors are static, (b) sensors are moving, but objects are static, (c) objects and sensors are moving. If a single object and a sensor family are in the environment and under constant motions, these three possibilities are equivalent. If motions are time dependent, or there are several objects and sensor families, multisensor fusion is a very complicated problem.

Dynamic multisensor fusion is temporal, for sensory data are integrated over time. If static multisensor fusion is considered as integration over space, dynamic fusion is double integration over space and time.

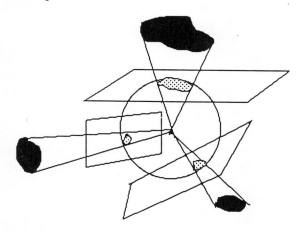

3. Spherical model

In the world space R^3, the sensor family is located at the origin O. In an usual vision system, a point $p(x,y,z)$ is projected to the image point $Q(X,Y)$ perspectively by the formulas $X=fx/z$ and $Y=fy/z$, where f (assumed to be 1 in the following) is the focal length and the z-axis is the viewing axis. The spherical model is to use the unit sphere as the image sphere on which all sensory data are represented. One improvement over the flat model is that we may represent sensory data of not only 360 degrees, but also the whole spherical solid angle.

The unit sphere is parametrized by two parameters θ and ϕ, where $0 \leq \theta \leq \pi$, $0 \leq \phi < 2\pi$. A point on the sphere is explicitly given by ($\cos\theta\cos\phi$, $\cos\theta \sin\phi$, $\sin\theta$). The angle θ is called the elevation angle and the angle ϕ is called the azimuth angle. The image point $Q(X,Y)$ is related to the spherical image point $S(\theta,\phi)$ by $X=\cot\theta\cos\phi$, $Y=\cot\theta \sin\phi$. The image intensity function is defined by the Lambertian model. Under the above relation, the intensity function $i(\theta,\phi)$ is $i(X(\theta,\phi),Y(\theta,\phi))$.

In the spherical coordinates (ρ,θ,ϕ) of R^3, the radial value (depth value) $\rho=\rho(\theta,\phi)$ is a function of

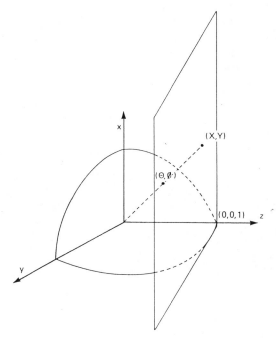

(θ,ϕ). As (θ,ϕ) varies, $\rho(\theta,\phi)$ sweeps out the object surface that is sensed. Thus, geometric information about the object surface can be determined by the radial value ρ and its derivatives $\partial\rho/\partial\theta$, $\partial\rho/\partial\phi$. In fact, $(1/\rho)(\partial\rho/\partial\theta)$ and $(1/\rho)(\partial\rho/\partial\phi)$ are the tangents of the slant and tilt angles. They determine the orientation at the point (ρ,θ,ϕ) on the object surface. By calculating second derivatives, the complete local surface geometry may be determined.

There are several 3-D data structures for sensory image understanding, robotics and spatial reasoning. We use the octree encoding scheme which represents an arbitrary object to a limited precision in a hierarchical tree structure. The higher resolution details are represented by the lower level nodes. For multisensor fusion, the spherical representation is more efficient than the

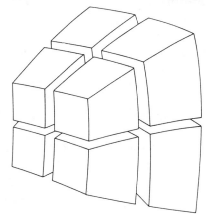

rectangular octree representation. The two are not exactly the same. The spherical representation is a viewer-centered system of spheres of varying radial values. An object is enclosed in a spherical sector, between two radial values. So, the octree structure centered about the object is curved. The usual rectangular octree is object-centered and the viewer is assumed at infinity, because of the orthographic projection.

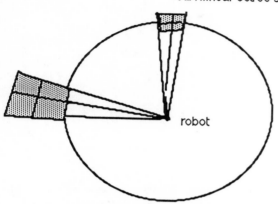

curvilinear octree structure

robot

4. Multisensor fusion techniques

The key problem of multisensor fusion is correspondence (or registration) of images. The correspondence of images of different sensor types is more difficult than the correlation of stereopsis and motion sequence. A generally accepted idea is to first reconstruct various object surfaces from images, and then integrate the object surfaces in the world space. Different kinds of sensory data require different algorithms and provide different kinds of image data. For instance, infrared sensing uses hot spots. Any object whose temperature is above a certain threshold radiates thermal energy. A moving tank generates a great deal of heat. A building or a machinery generate also heat. From different radiation profiles, we have different images. So thermal images are quite different from visual images of shading data. Thermal emitters are not necessarily good visual reflectors and visually bright objects may be dark in thermal scenes. However, in a 3-D object model, they provide the same kind of information about the object and thus can be integrated.

The parametrization $\rho=\rho(\theta,\phi)$ of the object surface is more realistic for perspectivity than the usual Monge patch parametrization $z=h(x,y)$ which is suitable for orthographic projection. The

spherical parametrization is further compatible with the spherical representation of octree structure. In the following sections, this parametrization will be used. Being the most common data, visual data is used for fusion with data of other kinds.

Range and visual data

The range data are represented by spatial points (ρ,θ,ϕ) in spherical coordinates. Geometric features of the 3-D approximating (Bezier or B-spline) surfaces may be calculated [4]. Two features - Gaussian and mean curvatures, characterize (the fundamental theorem of differential geometry) the local shape at each point. The range data are 3-D spatial information so that surface reconstruction is not needed. In addition to fusion of all spatial points from range data with the reconstructed surface points from visual sensing, the geometric features are used to guide the fusion.

In [5], we presented a photometric stereo method to reconstruct the local surface patch at a given point. We refer to the literature for other reconstruction methods. Given the reconstructed surface patch, we start to fuse the two surface patches. In general, there may not be any perfect point match between two surfaces. Thus, different averaging and approximating strategies may be adopted.

Infrared and visual data

Infrared imaging systems extend our vision beyond the short-wave length red into the far infrared by making visible the light naturally emitted by warm objects. While natural visual images are produced by reflection and by reflectivity differences, infrared images are produced by self emission and emissivity differences. The scene temperature produces the measured irradiance. The variations in the temperature correspond to the details in the visual scene, so an infrared image provides a visible analog of the infrared scene. The spatial patterns in the displayed infrared scene are similar to visual patterns. But there are significant differences.

Infrared scenes have fewer shadows than visible scenes and 3-D contour information is lost. But all surfaces radiate energy so that all forward surfaces tend to be visible. Infrared emitters are not necessarily good visual reflectors, thus visually bright objects may be dark in the infrared scenes and vice versa. One known approach is to build a knowledge base of infrared patterns for surface reconstruction. Another is treat reconstruction as an inverse heat conduction problem. So the temperature field is not the solution, but the unknown geometry of the object is. The geometry is solved as the output, while the temperature conditions in the forms of Dirichlet, Neumann, and Robin are assumed. This method requires numerical methods which are quite expensive. Therefore, it could be used in the preprocessing stage for building the knowledge base. In the two-dimensional case, we have the Laplace equation

$$\rho^2 \partial^2 T / \partial \rho^2 + \rho \partial T / \partial \rho + \partial^2 T / \partial \theta^2 = 0,$$

with various boundary conditions.

After the surface reconstruction from infrared images is performed, the 3-D information is merged with visual or range results in a similar way.

Radar and visual data

In radar sensing, shapes of aircraft, missiles, and satellites are determined by the radar cross section (RCS) method. RCS of an object characterizes the scattering properties of the object. The object is considered to be an ensemble of components, each of which can be geometrically approximated by a simple shape in such a way that RCS of the simple shape approximates that of the component it models. After calculation of RCSs, a proper combination of the component RCSs yields the estimate of RCS of the entire object. In radar sensing, the spherical model has been adopted in the near-zone to study wave fronts. In the far-zone, the flat model is used to study plane wave fronts.

The 3-D shape determined by radar data is again matched with the 3-D shape determined by visual, range or infrared images.

5. Conclusion

In this paper, several static multisensor fusion techniques are proposed. Experimentation of these techniques is under current research. Paper [1] and this paper form a two-part series of multisensor fusion.

References

1. S. Chen, Adaptive control of multisensor systems, this proceedings.
2. S. Chen, Multisensor fusion and navigation of mobile robots, International Journal of Intelligent Systems, Vol. 2, 1987, 227-252.
3. S. Chen, An intelligent computer vision system, International Journal of Intelligent Systems, Vol. 1, 1986, 15-28.
4. S. Chen and M. Penna, An efficient algorithm for calculating geometric features from range data, manuscript.
5. S. Chen and M. Penna, Motion analysis of deformable objects, Advances in Computer Vision and Image Processing, Vol. 3, Ed. T. Huang, JAL Press, 1987.

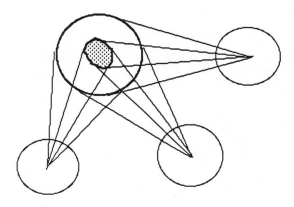

OPTIMAL AND SUBOPTIMAL DISTRIBUTED DECISION FUSION

Stelios C.A. Thomopoulos[1], Dimitri K. Bougoulias and Lei Zhang

Department of Electrical Engineering
Southern Illinois University
Carbondale, IL 62901

ABSTRACT

The problem of decision fusion in distributed sensors system is considered. A parallel sensor configuration is considered in which sensors monitor a common geographical volume and relay their decisions to a fusion center. The fusion center upon reception of the decision is responsible for fusing them into the final decision. Under conditional independence assumption, it is shown that the optimal test that maximizes the probability of detection for a fixed probability of false alarm consists of a Neyman-Pearson Test at the fusion and Likelihood-Ratio Tests st the sensors. Numerical evaluation of the optimal operating points is computationally intensive. Two computationally efficient suboptimal algorithms have been developed. Numerical results from extensive simulation in Rayleigh and Gaussian channels are presented.

INTRODUCTION

The problem of distributed decision fusion where a number of sensors transmit a compact form of information about a common observation space has attracted considerable attention recently, [1] through [7]. In this paper we consider the optimal decision scheme that maximizes the probability of detection at the fusion for fixed false alarm probability for the parallel sensor configuration, Fig. 1. According to this scheme, a number of sensors monitor the same geographical volume and transmit their decisions in regards with the nature of the true (binary) hypothesis to the fusion center which is responsible for combining the sensor decisions into a final one. We assume that the sensor decisions are independent conditioned on each hypothesis.

MAIN RESULTS

Theorem 1   The optimal decision rule for the distributed decision fusion problem consists of a Neyman-Pearson Test (N-P Test) at the fusion and Likelihood-Ratio Tests (L-R) at the sensors [8].

The proof of the theorem is based on the next two Lemmas which are quoted without proof below. The proofs of both Lemmas and the Theorem can be found in [8].

Lemma 1   Let the decisions u_k of the sensors be independent conditioned on each hypothesis. A necessary condition for a decision function $d(u_1, u_2, \ldots, u_N)$ to be threshold optimal is

$$d(A_k, U-A_k) = 1 \implies d(A_n, U - A_n) \geqq 1 \quad \text{if} \quad A_n > A_k.$$

where A_k is a set of decisions (sensors) favoring hypothesis H_1, and the set inequality is read in the lexicographic sense [8].

Lemma 2   The probability of detection P_{D_0} achieved at the fusion using the likelihood test

$$t(u_1, u_2, \ldots, u_N) \underset{H_0}{\overset{H_1}{\gtrless}} \lambda_0$$

is a monotonic increasing function of $P_{D_i} = Pr(u_i = 1 \mid H_1)$, $i = 1, 2, \ldots, N$ [8], where u_i designates the binary (0 or 1) decision of the i-th sensor.

Obtaining the optimal fusion rule numerically is a non-polynomial hard problem [9] if the optimal decision rule is searched among all the possible decision rules. From the set of all possible decision rules only the set of monotone increasing functions that depend on all the sensors qualify as optimal decision rules as the next Lemma states.

Lemma 3   The set of optimal decisions can be generated by considering only the monotone functions that depend on all the sensors and ignoring the monotone functions that depend on any subset of the sensors [8].

The number of all possible optimal decision rules depends on the number of sensors in the system. It can be compute recursively using Theorem 2, [9]. An indication of the non-polynomial complexity of the problem can be obtained by referring to Tables I and II.

---

1 This research is sponsored by the SDIO/IST and managed by the Office of Navl Research under Contract N00014-k-0515.

NUMERICAL SOLUTIONS TO THE FUSION PROBLEM

In [6] the optimal combining rule in a parallel sensor configuration was given in terms of a set of coupled, nonlinear equations whose solution depends on the decision rule and cannot be solved in general. Furthermore, they exhibit numerical problems related to the Lagrangian method which was used to derive them [?]. Two suboptimal algorithms that allow the determination of the decision rule have been developed. The two algorithms allow the determination of a fusion rule using a one dimensional minimization and a one dimensional search, and are computationally very efficient. The algorithms are based on the sequential optimization of the Lagrangian w.r.t. the different sensors assuming that the thresholds of previously optimized sensors are set so that the sensors operate at either zero or one probability of detection. The two algorithms will be referred as SOFA 1 and SOFA 2 respectively and are presented next.

SOFA 1 ALGORITHM: Let 1, 2, ..., N be an arbitrary ordering of N sensors. Starting from the N-th sensor, the threshold of the k-th sensor as determined by SOFA 1 is given by

$$\lambda_k = \lambda_0 \frac{C_0^{N,N-1,\ldots,k}}{C_1^{N,N-1,\ldots,k}} \tag{1}$$

where

$$C_i^{N,N-1,\ldots,k} = \sum_{U_{N,N-1,\ldots,k}} [d(0,0,\ldots,0,1,U_{N,N-1,\ldots,k}) - d(0,0,\ldots,0,0,U_{N,N-1,k})]p(U_{N,N-1,\ldots,k}|H_i) , i=0,1 \tag{2}$$

where λ_k designates the threshold of the k-th sensor, λ_0 is the threshold at the fusion,

$$d(u_1, u_2, \ldots, u_N) = Pr(u_0 = 1 \mid u_1, u_2, \ldots, u_N) \tag{3}$$

is the decision function at the fusion center with u_i designating the binary decision of the i-th sensor and u_0 the decision at the fusion, H_i, i = 1, 0 is the true hypothesis and the alternative, and $U_{N,N-1,\ldots,k}$ is the set of decisions of all the sensors excluding those (decisions) of the N, N-1, ..., k sensors whose thresholds have already been determined. Furthermore, for the first sensor

$$\lambda_1 = \lambda_0 \tag{4}$$

where λ_0 is the threshold at the fusion center.

SOFA 2 ALGORITHM: Let 1, 2, ..., N be an arbitrary ordering of N sensors. Starting from the N-th sensor, the threshold of the k-th sensor as determined by SOFA 2 is given by

$$\lambda_k = \lambda_0 \frac{D_0^{N,N-1,\ldots,k}}{D_1^{N,N-1,\ldots,k}} \tag{5}$$

where

$$D_i^{N,N-1,\ldots,k} = \sum_{U_{N,N-1,\ldots,k}} [d(1,1,\ldots,1,1,U_{N,N-1,\ldots,k}) - d(0,0,\ldots,0,0,U_{N,N-1,k})]p(U_{N,N-1,\ldots,k}|H_i) , i = 0,1 \tag{6}$$

where λ_k designates the threshold of the k-th sensor, λ_0 is the threshold at the fusion,

$$d(u_1, u_2, \ldots, u_N) = Pr(u_0 = 1 \mid u_1, u_2, \ldots, u_N) \tag{7}$$

is the decision function at the fusion center with u_i designating the binary decision of the i-th sensor and u_0 the decision at the fusion, H_i, i = 1, 0 is the true hypothesis and the alternative, and $U_{N,N-1,\ldots,k}$ is the set of decisions of all the sensors excluding those (decisions) of the N, N-1, ..., k sensors whose thresholds have already been determined. Furthermore, for the first sensor

$$\lambda_1 = \lambda_0 \tag{8}$$

where λ_0 is the threshold at the fusion center.

The derivation of the equations that define the two algorithms can be found in [7].

NUMERICAL TESTING

Several numerical results from the application of the two algorithms in distributed decision fusion with various numbers of sensors are given and the performance of the algorithms is compared with the globally optimal solution obtained by direct optimization. The two algorithms were tested in slow-fading Rayleigh channels and in Gaussian channels. For the slow-fading Rayleigh channel the probability of false alarm and probability of detection for the kth sensor are given by:

$$P_{F_k} = [\lambda_k(1 + \epsilon_k)]^{-(1 + \frac{1}{\epsilon_k})} \quad \text{and} \quad P_{D_k} = P_{F_k}^{(\frac{1}{1+\epsilon_k})} \tag{9}$$

where P_{F_k} and P_{D_k} are the probabilities of false alarm and probabilities of detection at the sensor, λ_k is the threshold at the kth sensor, and ϵ_k represents the signal-to-noise ratio at the kth sensor [9].

For the gaussian case the relationship between the probabilities of false alarm, involve the error function and depend on the signal-to-noise ratio. However, no close form relationship between the false alarm probability and the probability of detection can be obtained [10].

For the two different channels numerical results have been obtained for two, three and four sensors. Figures 2, 3 and 4 were obtained from a fusion system with two, three, and four identical sensors respectively operating in a slow fading Rayleigh channel using SOFA 1. The decision rules that were tested were the OR, AND and the Majority Logic (ML) when possible. The optimal solution for the three policies was also obtained with direct multidimensional minimization. SOFA 1's results were found to be extremely close to the optimal ones for the OR and AND decision rules in all three cases, and sufficiently close in the case of the ML. Figure 5 was obtained using SOFA 1 in a fusion system with three sensors, two of them operating at the same signal-to-noise ratio (SNR) and one at a different one, and decision rule OR. The solution with SOFA 1 was found to be identical to the optimal one again. Similar conclusions were drawn from numerical results obtained using SOFA 2.

Figures 6, 7 and 8 were obtained from a fusion system with two, three, and four identical sensors respectively operating in a channel with additive Gaussian noise using SOFA 1. The decision rules that were tested were the OR, AND and the Majority Logic (ML) when possible. The optimal solution for the three policies was also obtained with direct multidimensional minimization. In the case of two sensors, it was found that SOFA 1 did not track the optimal solution close enough under either policy, Fig. 6. However, in the case of three and four sensors it was found that SOFA 1 was able to track the optimal solution in the case of the OR decision rule very close, Figs. 7 and 8. In the case of AND, which turned out to be the overall optimal decision rule in the range of the SNR's that was tested, SOFA 1 was still able to track the optimal solution fairly close. The biggest discrepancy between the SOFA 1 solution and the optimal one occurred under the ML rule. Furthermore, in the case of three dissimilar sensors, SOFA 2 was able to track the optimal solution very closely in the case of the OR rule, Fig. 9, but failed to track closely the optimal solution in the case of the AND rule for low SNR, Fig. 10. The performance of SOFA 2 in the Gaussian case was found to be inferior to that of SOFA 1 in the cases of similar sensors under the AND policy. However, SOFA 2 produced results that were closer to the optimal ones in the case of dissimilar sensors. The results from the Gaussian case are inconclussive and further investigation and testing is required.

CONCLUSION

The optimal fusion rule for the distributed sensor fusion problem of Figure 1 was discussed. Numerical results from two computationally efficient algorithms were presented fro slow fading Rayleigh and Gaussian channels. It was found that the algorithms perform nearly optimally in the case of the Rayleigh channel. In the Gaussian case, the algorithms were not always able to track very accurately the optimal solution in all the tested decision rules. Further investigation and testing of the two algorithms is required to assess their merits and possible limitations.

REFERENCES
[1] Tenney, R. R. and Sandell, N.R., Jr., "Detection with Distributed Sensors," IEEE Trans. on Aerospace and Electronic Systems, Vol. AES-17, July 1981, pp. 501-510.
[2] Sadjadi, F. A., "Hypothesis Testing in A Distributed Environment," IEEE Trans. on Aerospace and Electronic Systems, Vol. AES-22, March 1986, pp. 134-137.
[3] Chair, Z. and Varshney, P. K., "Optimal Data Fusion in Multiple Sensor Detection Systems," IEEE Trans. on Aerospace and Electronic Systems, Vol. AES-22, No. 1, January 1986, pp. 98-101.
[4] Thomopoulos, S. C. A., Viswanathan, R. and Bougoulias, D. K., "Optimal Decision Fusion in Multiple Sensor Systems," Proceedings of the 24th Allerton Conference, October 1-3, 1986, Allerton House, Monticello, Illinois, pp. 984-993.
[5] Thomopoulos, S. C. A., Viswanathan, R. and Bougoulias, D. K., "Optimal Decision Fusion in Multiple Sensor Systems," IEEE Trans. on Aerospace and Electronic Systems, Vol. AES-23, No. 5, Sept. 1987.
[6] Srinivasan, R., "Distributed Radar Detection Theory," IEE Proceedings, Vol. 133, Pt.F, No. 1, February 1986, pp. 55-60.
[7] Thomopoulos, S. C. A., Viswanathan, R. and Bougoulias, D.K., "Optimal and Suboptimal Distributed Decision Fusion," Technical Report, TR-SIU-EE-87-5, Aug. 1987. Also submitted to IEEE Trans. on Aerospace and Electronic Systems.
[8] Thomopoulos, S. C. A., Viswanathan, R. and Bougoulias, D.K., "Optimal and Suboptimal Distributed Decision Fusion: The N-P/L-T Test," 22nd Annual Conference on Information Sciences and Systems, Princeton University, March 16-18, 1988, to appear.
[9] DiFranco, J.V. and Rubin, W.L., "Radar Detection," Artech House, Inc.,Dedham, MA, 1980.
[10] Van Trees, H.L., "Detection, Estimation and Modulation Theory," John Wiley & Sons, NY, 1968.

Table I

Number of Monotone Increasing Functions and % Reduction

Number of Sensors N	Number of Monotone Functions	Number of all Possible 2^{2^N} Functions	Percentage Reduction %
1	3	4	25
2	6	16	62.5
3	20	256	92.19
4	168	65,536	99.74
5	7,581	4.2949673×10^9	99.99982
6	7,828,354	1.8446744×10^{19}	100

Table II

Total Number of Functions Searched for the Set of Optimal Thresholds

Number of Sensors N	L_N (= # of Monotone Functions -2)	Total Number of Functions R_N	Percentage Reduction %
1	1	1	0.00
2	4	2	50.00
3	18	9	50.00
4	166	114	31.13
5	7,579	6,894	9.03
6	7,828,352	7,786,338	0.54

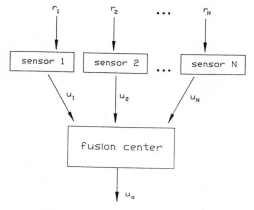

FIGURE 1   Distributed Sensor Fusion System

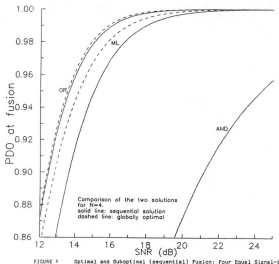

Comparison of the two solutions for N=2.
solid line: sequential solution
dashed line: globally optimal

FIGURE 2   Optimal and Suboptimal (sequential) Fusion: Two Equal Signal-to-Noise Ratio Sensors. Distribution: Rayleigh. Algorithm: SOFA 1. Probability of False Alarm at Fusion $P_F = 10^{-6}$.

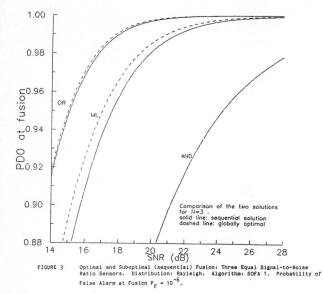

Comparison of the two solutions for N=3.
solid line: sequential solution
dashed line: globally optimal

FIGURE 3   Optimal and Suboptimal (sequential) Fusion: Three Equal Signal-to-Noise Ratio Sensors. Distribution: Rayleigh. Algorithm: SOFA 1. Probability of False Alarm at Fusion $P_F = 10^{-6}$.

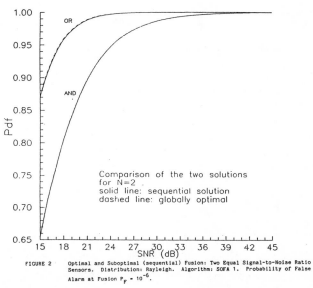

Comparison of the two solutions for N=4.
solid line: sequential solution
dashed line: globally optimal

FIGURE 4   Optimal and Suboptimal (sequential) Fusion: Four Equal Signal-to-Noise Ratio Sensors. Distribution: Rayleigh. Algorithm: SOFA 1. Probability of False Alarm at Fusion $P_F = 10^{-6}$.

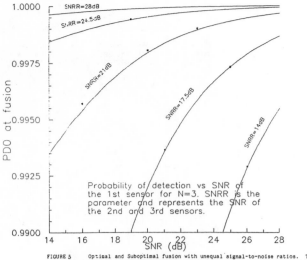

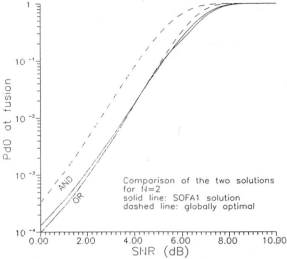

FIGURE 5   Optimal and Suboptimal fusion with unequal signal-to-noise ratios. Three
sensors.   The optimal solution is designated by *.   Algorithm: SOFA 1.
Probability of False Alarm at Fusion $P_F = 10^{-6}$.

Figure 6   Optimal and SOFA 1 solutions for two equal SNR sensor fusion in
Gaussian channel. False Alarm Probability = 10^{-6}.

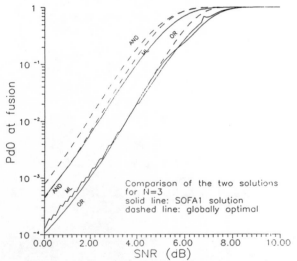

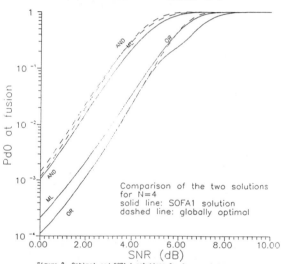

Figure 7   Optimal and SOFA 1 solutions for three equal SNR sensor fusion in
Gaussian channel. False Alarm Probability = 10^{-6}.

Figure 8   Optimal and SOFA 1 solutions for four equal SNR sensor fusion in
Gaussian channel. False Alarm Probability = 10^{-6}.

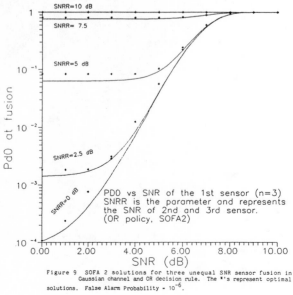

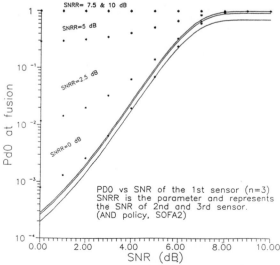

Figure 9   SOFA 2 solutions for three unequal SNR sensor fusion in
Gaussian channel and OR decision rule.   The *'s represent optimal
solutions. False Alarm Probability = 10^{-6}.

Figure 10 SOFA 2 solutions for three unequal SNR sensor fusion in Gaussian
channel and AND decision rule.   The *'s represent optimal
solutions.False Alarm Probability = 10^{-6}.

DISTRIBUTED DETECTION WITH CONSULTING SENSORS AND COMMUNICATION COST

Stelios C.A. Thomopoulos[1] and Nickens N. Okello

Department of Electrical Engineering
Southern Illinois University
Carbondale, IL 62901

ABSTRACT

The problem of distributed detection with consulting sensors is formulated and solved when there is a communication cost associated with any exchange of information (consultation) between the sesnors. We consider a system of two sensors, S1 and S2, in which S1 is the primary sensor responsible for the final decision u_0, while S2 is a consulting sensor capable of relaying its decision u_2 to S1 when requested by S1. In the scenario that is considered, the final decicion u_0 is based either on the raw data available to S1 only, or, it may, under certain request conditions, also take into account the decision u_2 of sensor S2. Random and non-random request schemes have been analysed and numerical results for both request schemes are presented for a slowly fading Rayleigh channel.

I. INTRODUCTION

Considerable research has been focused lately on the problem of distributed decision fusion [1-6] when a number of distributed sesnors receive data from a commom volume, come up with a first-stage decision, and then transmit their decisions to a fusion center which comes up with the final decision by fusing the sensor decisions (or, some form of compact inforamtion received from the sensors). The main assumption in the bulk of the related literature is that transmission of information from the sensor to the fusion (and, maybe, the opposite way) is done at no cost. This implies that exchange of information between the sensors and the fusion is possible at rates limited only by the physical bounds of the channel capacity. The main emphasis is given then in determining the optimal sensor configuration (parallel, serial, or combination) [5-6] and, the fusion logic (AND, OR, etc) for an array of sensors [5-6].

In this paper we consider the problem of distributed decision making with two consulting sensors in which every inter-sensor communication incurs some risk, thus making continuous sensor communication a very expensive and prohibitive proposition. We are interested in determining the optimum detection scheme given that a certain amount of risk (or, communications cost) can be tolerated. Furthermore, we are interested in the optimal decision rule that minimizes cost functionals that involve the probability of false alarm, the communication cost and the probability of miss. Different formulations are possible and four are being discussed in this paper.

The problem of team-decision with risk is motivated by Communication-Command and Control - C^3 - applications [7]. However, practical application areas for this aspect of distributed detection are not confined to C^3 problems but extend to other fields, like medical diagnosis, cryptography, etc., where team decision problems exist and consultation takes place always at a fee (cost).

II. TEAM DECISION SCHEMES

The team-decision scenario that we analyze in this paper consists of a dual-sensor system and binary hypothesis testing. Extension to multiple hypothesis testing is straight forward. Due to bandwidth limitations and the sensitivity of the data, no transmission of raw data between the two sensors is allowed. The sensors exchange only request signals and decisions with, may be, some additional quality information bits, like the degree of confidence associated with their decisions. For reasons of analytical clarity and presentation convinience, we present analytical results only for the cases where the primary sensor transmits request signals to the consultant sesnor, whereas the latter relays only its binary decisions back to the primary. With these assumptions the class of scenaria that fit into our analysis have as follows.

Consider a system of two sensors S1 and S2, Fig.1, in which S1 acts as the primary sensor and decides (based on locally available raw data) in favor of either hypothesis H_0 or hypothesis H_1 (constituting the final decision), or, declares ignorance, I, when its decision falls in the uncertainty region (See Fig. 1 and below for the definition of the uncertainty region). When ignorance is declared, a request message is transmitted from S1 to S2, accompanied, if necessary, by quality information regarding S1's assessmest as to

1. This research is sponsored by the SDIO/IST and managed by the Office of
   Naval Research under Contract N00014-k-0515.

what hypothesis is most probable and the degree of confidence S1 attaches to it. Sensor S2 then analyses its own raw data (taking into account the fact that it has been consulted and any additional information relayed by S1) and arrives at a decision u_2 in favor of either H_0 or H_1. The decision u_2 is then transmitted to S1 which treats it either as the final decision u_0 or, reprocesses it along with its own original data to produce a final decision u_0. Another possibility is to include a third option of indecision for sensor S2 in which case S1 must then reprocess its original data together with u_2. While this latter arrangement may improve the overall system performance by some, the associated analytical complexity renders it unattractive.

The dual-sensor decision making problem is formulated as follows. Given a dual-sensor configuration, where the primary sesnor makes a preliminary decision conserning the nature of the true (binary) hypothesis and, either decides in favor of one the two hypotheses, in which case the decision is final, or, decides to consult the consultant sensor if its data falls in the indecision region, what is the optimal test that optimizes some joint performance criterion if, the primary sensor sends only request bits to the consultant and the consultant arrives at a decision based on its own data and the fact that the primary has declared ignorance (implicity induced by the fact that it (the consultant) has been consulted), and then relays its decision back to the primary, which considers it as the final decision or, reprocess it along with its own data.

In the analysis that follows we assume that the decisions of the primary and consulting sensors are mutually independent conditioned on each hypothesis.

III. DIFFERENT OPERATING SCHEMES

Given a dual-sensor systems with primary sensor S1 and consulting sensor S2, and the assumptions of the previous section, the following operating schemes are considered.

III.1   Random Consultation With Fixed Probability.

S1 randomly consults S2 with a fixed probability of request P_R in order to achieve a specified miss probability P_M for fixed probability of false alarm P_F. In this scenario we assume that when S2 is consulted, it relays its decision to S1, which, in turns, combines it with its raw data in order to come up with the final decision. The distinguishing feature of this scheme is that the decision to consult is random and is made independently of the degree of confidence that sensor S1 may have on its initial decision u_1. The major advantages of the scheme is that: a) it is simple to analyze; and b) its performance does not depend on the prior probabilities of the two hypotheses which may, very often, be unknown in C^3 and other applications.

The scheme of random consultation is equivalent to choosing P_R such that, for a fixed false alarm probability α_0, the system operates at a point between the ROC (Receiver Operating Characteristic) of S1 alone [8], and the ROC of the serial combination of S1 and S2 [6] for the same probability of false alarm, Fig. 2. For both cases, the optimal test that minimizes the probability of miss for fixed probability of false alarm is the Neyman-Pearson (N-P) Test [6]. Hence, for a fixed false alarm probability α_0, the miss probability P_M is

$$P_M = P_{MS1}(1-P_R) + P_{MS12} P_R \tag{1}$$

where P_{MS1} and P_{MS12} designate the miss probablities of S1 operating alone and of the serial combination of S1 and S2 [6] respectively. Solving for P_R we obtain

$$P_R = \frac{P_M - P_{MS1}}{P_{MS12} - P_{S1}} \tag{2}$$

To derive the ROC of the serial combination of S1 and S2, we consider a system of two sensors S1 and S2 in which the decision u_2 of sensor S2 is transmitted to sensor S1 and is then used together with the raw data Z_1 available to S1 to arrive at a final decision u_1. To that extend, we follow an analysis similar to [6]. Denoting the distribution of Z_1 as $p(Z_1|H_0)$ and $p(Z_1|H_1)$, the likelihood ratio at sensor S1 becomes

$$\frac{L(r_1,u_2|H_1)}{L(r_1,u_2|H_0)} = \frac{p(r_1|H_1)[P_{D2}\delta(u_2-1)+(1-P_{D2})\delta(u_2-1)]}{p(r_1|H_0)[P_{F2}\delta(u_2-1)+(1-P_{F2})\delta(u_2-1)]} \tag{3}$$

where,

$$P_{D2} = Pr(u_2=1|H_1)$$

$$\tag{4}$$

$$P_{F2} = Pr(u_2=1|H_0)$$

are the detection and false alarm probabilities at S2 respectively, $u_2=k$ implies that sensor S2 decides H_k, $k=0,1$ and, $\delta(x)$ is the Kronecker's delta

$$\delta(x) = \begin{vmatrix} 1 & x=0 \\ 0 & x \neq 0 \end{vmatrix}$$

Hence, if t is the threshold at sensor S1, the test at S1 reduces to

$$\frac{p(r_1|H_1)\,P_{D2}}{p(r_1|H_0)\,P_{F2}} \overset{H_1}{\underset{H_0}{\gtrless}} t \qquad \text{if } u_2=1$$

$$\tag{5}$$

$$\frac{p(r_1|H_1)(1-P_{D2})}{p(r_1|H_0)(1-P_{F2})} \overset{H_1}{\underset{H_0}{\gtrless}} t \qquad \text{if } u_2=0$$

Alternatively,

$$\Lambda(r_1) \overset{H_1}{\underset{H_0}{\gtrless}} \begin{array}{ll} t_{1,1} & \text{if } u_2=1 \\ t_{1,0} & \text{if } u_2=0 \end{array} \tag{6}$$

where $\Lambda(r_1) = \dfrac{p(r_1|H_1)}{p(r_2|H_0)}$

and

$$\frac{t_{1,1}}{t_{1,0}} = \frac{P_{F2}(1-P_{D2})}{P_{D2}(1-P_{F2})} \tag{7}$$

If the sensors S1 and S2 operate in a slowly fading Rayleigh enviroment, the false alarm and detection probabilities for S1 given $u_2=1$ are [9]

$$P_{F1,1} = [t_{1,1}(1+\varepsilon 1)]^{-(1+\varepsilon 1)/\varepsilon 1} \tag{8}$$

$$P_{D1,1} = [P_{F1,1}]^{1/(1+\varepsilon 1)} \tag{9}$$

and the corresponding expressions for $u_2=0$ are

$$P_{F1,0} = [t_{1,0}(1+\varepsilon 1)]^{-(1+\varepsilon 1)/\varepsilon 1} \tag{10}$$

$$P_{D1,0} = [P_{F1,0}]^{1/(1+\varepsilon 1)} \tag{11}$$

where $\varepsilon 1$ is the signal-to-noise ratio (SNR) at S1.

Thus, the overall false alarm probability P_F of the dual-sensor system is given by

$$\begin{aligned} P_F = &\Pr(\Lambda(r_1) > t_{1,0} \mid H_0, u_2=0)\,\Pr(u_2=0|H_0) \\ &+ \Pr(\Lambda(r_1) > t_{1,1} \mid H_0, u_2=1)\,\Pr(u_2=1|H_0) \end{aligned} \tag{12}$$

or,

$$\begin{aligned} P_F = &\Pr(\Lambda(r_1) > t_{1,0} \mid H_0)\,\Pr(u_2=0 \mid H_0) \\ &+ \Pr(\Lambda(r_1) > t_{1,1} \mid H_0)\,\Pr(u_2=1 \mid H_0) \end{aligned} \tag{13}$$

or,

$$P_F = P_{F1,0}(1-P_{F2}) + P_{F1,1}\,P_{F2} \tag{14}$$

Similarly, the overall detection probability P_D is given by

$$\begin{aligned} P_D = &\Pr(\Lambda(r_1) > t_{1,0} \mid H_1)\,\Pr(u_2=0 \mid H_1) \\ &+ \Pr(\Lambda(r_1) > t_{1,1} \mid H_1)\,\Pr(u_2=1 \mid H_1) \end{aligned} \tag{15}$$

or,

$$P_D = P_{D1,0}(1-P_{D2}) + P_{D1,1}\,P_{D2} \tag{16}$$

If the dual-sensor system is constrained to operate at a false alarm probability α_0, then equation (14) gives

$$\alpha_0 = P_{F1,0}(1-P_{F2}) + P_{F1,1}\,P_{F2} \tag{17}$$

From equations (7) and (8) it follows that

$$P_{F1,1} = \left[t_{1,0}\,\frac{P_{F2}(1-P_{D2})}{P_{D2}(1-P_{F2})}(1+\varepsilon 1) \right]^{-(1+\varepsilon 1)/\varepsilon 1} \tag{18}$$

Substituting $t_{1,0}$ in (18) from (4) and noticing that, for a slowly fading Rayleigh enviroment,

$$P_D = (P_{F2})^{1,9\,(1+\varepsilon 2)} \tag{19}$$

where $\varepsilon 2$ is the SNR at S2, (18) yields an expression for $P_{F1,1}$ in the form

$$P_{F1,1} = f(P_{F1,0}, P_{F2}) \tag{20}$$

From equations (17) and (20), the constraint (17) becomes

$$\alpha_0 = P_{F1,0}(1-P_{F2}) + P_{F2}\,f(P_{F1,0}, P_{F2})$$

or,

$$\alpha_0 = g(P_{F1,0}, P_{F2}) \tag{21}$$

Using (9), (11) and (21) the overall detection probability P_D in (16) can be expressed as

$$P_D = (P_{F1,0})^{1/(1+\varepsilon1)}[1 - (P_{F2})^{1/(1+\varepsilon2)}]$$

$$+ (P_{F1,1})^{1/(1+\varepsilon1)} (P_{F2})^{1/(1+\varepsilon2)} \qquad (22)$$

or,

$$P_D = (P_{F1,0})^{1/(1+\varepsilon1)} [1-(P_{F2})^{1/(1+\varepsilon2)}]$$
$$+ [g(P_{F2}, P_{F1,0})]^{1/(1+\varepsilon2)} (P_{F2})^{1/(1+\varepsilon1)} \qquad (23)$$

or,

$$P_D = h(P_{F1,0}, P_{F2}) \qquad (24)$$

From equations (19) and (22), it follows that the optimal test for the dual-sensor system can be obtained by solving the following maximization problem:

Maximize $P_D = h(P_{F1,0}, P_{F,2})$
Subject to $\alpha_0 = g(P_{F1,0}, P_{F,2})$ $\qquad (25)$
where $0 \leq P_{F1,0} \leq 1$ and $0 \leq P_{F,2} \leq 1$.

If P_{M12} denotes the optimum miss probability for the serial arrangement between sensors S1 and S2 [6], and P_M is the miss probability for S1 operating alone, then for a fixed false alarm probability α_0, the miss probability P_M of the dual-sensor system under random consultation is given by

$$P_M = P_{M1}(1 - \beta_0) + P_{M12} \beta_0 \qquad (26)$$

where β_0 is the probability level at which sensor S1 randomly consults S2, i.e. P_R, and P_{M12} is the miss probability for the serial sensor configuration with continuous transmission of the decision of S2 to S1 [6]. For the slowly fading Rayleigh channel,

$$P_{M1} = [1 - (\alpha_0)^{1/(1+\varepsilon1)}] \qquad (27)$$

Hence, the overall detection probability P_D subject to false alarm probability α_0 and request probability β_0 is given by

$$P_D = 1 - [1 - (\alpha_0)^{1/(1+\varepsilon1)}][1-\beta_0] - P_{M12} \beta_0 \qquad (28)$$

Note that because P_{M12} is not a function of the request probability β_0, the values of $P_{F1,0}$ and $P_{F,2}$ corresponding to the optimum solution will only change with α_0 but not with β_0. Numerical results of the performance of this scheme are given in Figure 3.1.(a)   . From these figures it is seen that the performance of the random request scheme is monotonic w.r.t. the SNR at the sensors, as expected. (For all the numerical examples throughout this paper it was assumed that both sensors were identical and operated at the same SNR.) Opposite to the schemes that are discussed next, the performance of this scheme does not depend on the prior probabilities, its major advantage besides its simplicity. However, it will be seen next, that its performance is far inferior to the schemes that are discussed next.

III.2   Non-random Consultation Schemes.

In non-random consultation schemes the decision to consult is made only when the initial decision u_1 of S1 falls within the indecision region, Fig. 2, otherwise, if u_1 falls outside the region of indecision, it is taken as final.

While several different formulations are possible, we shall only concern ourselves with the case in which S1 may consult S2 but does not relay any quality information regarding its initial findings. In addition, S2, upon being requested, processes its own raw data taking into account the fact that it has been consulted, and transmits it decision, u_2, to S1 which then treats it as final.

Consider a system of two sensors S1 and S2 with conditional likelihood ratio distributions as in Figures 1. Assume that S1 has an uncertainty region (t', t_1'') such that, if $\Lambda_1(r_1) \varepsilon(t_1', t_1'')$, it consults S2, without transmitting any quality information about its preliminary decision to S2. S2 then processes its data conditioned on the the event that S1's decision falls in the indecision region I, induced by the fact that it has been consulted, and then relays its decision u_2 to S1 which takes it as final for the entire system. It, therefore, follows that S1 decides according to the following scheme:

$$\Lambda_1(r_1) < t_1' \quad : \quad \text{choose } H_0$$
$$t_1' \le \Lambda_1(r_1) \le t_1'' \quad : \quad \text{choose I (Ignorance)} \tag{29}$$
$$\Lambda_1(r_1) > t_1'' \quad : \quad \text{choose } H_1$$

while S2 employs the familiar likelihood ratio test given by:

$$\Lambda_2(r_2, u_1=I) \underset{H_0}{\overset{H_1}{\gtrless}} t_2 \tag{30}$$

If u_0 denotes the final decision of the system then the overall miss probability P_M is given by:

$$P_M = P(u_0=0|H_1)$$
$$= \sum_{u_1} P(u_0=0|u_1,H_1) \, P(u_1|H_1)$$
$$= P(u_0=0|u_1=1,H_1) \, P(u_1=1|H_1) + P(u_0=0|u_1=0,H_1) \, P(u_1=0|H_1)$$
$$+ P(u_0=0|u_1=I,H_1) \, P(u_1=I|H_1) \tag{31}$$

The first part of the left-hand side of (31) equals zero, while the second and the third parts can be simplified to give:

$$P_M = P(u_1=0|H_1) + P(u_2=0|u_1=I,H_1) \, P(u_1=I|H_1) \tag{32}$$

Expressing P_M in terms of the likelihood ratio $\Lambda_1(r_1)$ and $\Lambda_2(r_2,I)$, we get

$$P_M = \int_{\Lambda_1<t_1''} dP(\Lambda_1|H_1) + \int_{\Lambda_2<t_2} dP(\Lambda_2|H_1) \int_{t_1'}^{t_1''} dP(\Lambda_1|H_0) \tag{33}$$

Similarly, the overall probability of false alarm is given by

$$P_F = \int_{\Lambda>t_1''} dP(\Lambda_1|H_0) + \int_{\Lambda_2>t_2} dP(\Lambda_2|H_0) \int_{t_1'}^{t_1''} dP(\Lambda_1|H_0) \tag{34}$$

and the probability of request

$$P_R = \int_{t_1'}^{t_1''} dP(\Lambda_1) = \int_{t_1'}^{t_1''} [\, dP(\Lambda_1|H_0)P_0 + dP(\Lambda_1|H_1)(1-P_0)] \tag{35}$$

Note that it is necessary to express the likelihood ratio $\Lambda_2(r_2,u_1=I)$ in terms of $\Lambda_2(r_2)$ in order to be able to evaluate the integrals

$$\int_{\Lambda_2>t_2} dP(\Lambda_2|H_0) \quad \text{and} \quad \int_{\Lambda_2>t_2} dP(\Lambda_2|H_1) \tag{36}$$

By definition,

$$\Lambda_2(r_2, u_1=I) = \frac{P(r_2, u_1=I|H_1)}{P(r_2, u_1=I|H_1)} \underset{H_0}{\overset{H_1}{\gtrless}} t_2 \tag{37}$$

$$= \frac{P(r_2,H_1) \, P(u_1=I|H_1)}{P(r_2,H_0) \, P(u_1=I|H_0)} \underset{H_0}{\overset{H_1}{\gtrless}} t_2$$

or,

$$\Lambda_2(r_2,u_1=I) = \Lambda_2(r_2) \frac{\int_{t_1'}^{t_1''} dP(\Lambda_1|H_1)}{\int_{t'}^{t''} dP(\Lambda_1|H_0)} \underset{H_0}{\overset{H_1}{\gtrless}} t_2 \tag{38}$$

Hence,

$$\Lambda_2(r_2) \underset{H_0}{\overset{H_1}{\gtrless}} t_2 \frac{\int_{t_1'}^{t_1''} dP(\Lambda_1|H_0)}{\int_{t_1'}^{t_1''} dP(\Lambda_1|H_1)} \hat{=} t_2' \tag{39}$$

It, therefore, follows that

$$\int_{\Lambda_2>t_2} dP(\Lambda_2(r_2,u_1=I)|H_i) = \int_{\Lambda_2>t_2'} dP(\Lambda_2(r_2)|H_i) \, , \quad \text{for } i = 0, 1. \tag{40}$$

Thus,

$$P_M = \int_{\Lambda_2<t_1'} dP(\Lambda_1|H_1) + \int_{\Lambda_2<t_2} dP(\Lambda_2(r_2)|H_1) \int_{t_1'}^{t_1''} dP(\Lambda_1|H_0) \tag{41}$$

Similarly,

$$P_F = \int_{\Lambda_1 > t_1''} dP(\Lambda_1 | H_o) + \int_{\Lambda_2 > t_2} dP(\Lambda_2(r_2) | H_o) \int_{t_1'}^{t_1''} dP(\Lambda_1 | H_o) \tag{42}$$

and

$$P_R = \int_{t_1'}^{t_1''} dP(\Lambda_1) = \int_{t_1'}^{t_1''} [\, dP(\Lambda_1 | H_o) \, P_o + dP(\Lambda_1 | H_1) \, (1-P_o) \,] \tag{43}$$

For the case of a slowly fading Rayleigh channel, if the threshold of the sensor is t and the SNR is ε, P_F and P_D are given by [9]

$$P_F(t) = [t(1+\varepsilon)]^{-1\frac{1}{\varepsilon}} = \int_{\Lambda > t} dP(\Lambda | H_o) \tag{44}$$

$$P_D(t) = [P_F(t)]^{1/(1+\varepsilon)} \tag{45}$$

and

$$P_M(t) = 1 - P_D(t) = 1 - [P_F(t)]^{1/(1+\varepsilon)} = \int_{\Lambda < t} dP(\Lambda | H_1) \tag{46}$$

where the product $t(1+\varepsilon) \geq 1$ to ensure that P_D and $P_F \leq 1$.

If S1 operates at ε_1 SNR, and S2 operates at ε_2, from (33), (34) and (35) it follows

$$P_M = P_M(t_1') + P_M(t_2') [P_M(t_1'') - P_M(t_1')] \tag{47}$$

or,

$$P_M = 1.0 - [P_F(t_1')]^{1/(1+\varepsilon_1)} + [1-(P_F(t_2'))^{1/(1+\varepsilon_2)}][(P_F(t_1'))^{1/(1+\varepsilon_1)} - (P_F(t_1''))^{1/(1+\varepsilon_1)}] \tag{48}$$

Similarly,

$$P_F = P_F(t_1'') + P_F(t_2')[P_F(t_1') - P_F(t_1'')] \tag{49}$$

and

$$P_R = P_o[P_F(t_1') - P_F(t_1'')] + (1-P_o)[P_M(t_1'')-P_M(t_1')] \tag{50}$$

or,

$$P_R = P_o[P_F(t_1') - P_F(t_1'')] + (1 - P_o)[(P_F(t_1'))^{1/(1+\varepsilon_1)} - (P_F(t_1''))^{1/(1+\varepsilon_1)}] \tag{51}$$

Note that the expressions for P_R, P_F and P_M are subject to the constraint $t_1'' \geq t_1'$ which in turn implies that

$$P_F(t_1') \geq P_F(t_1'') \quad \text{and} \quad P_M(t_1') \leq P_M(t_1'') \tag{52}$$

Using equations ((48), (49) and (51), it is possible to determine the optimum thresholds t_1', t''_1 and t_2' iteratively for a wide range of formulations.

IV. DIFFERENT FORMULATIONS FOR NON-RANDOM REQUEST SCHEMES

A number of possible formulations may be arrived at depending on the specific application and objective. We shall, however, only concern ourselves with the following three different formulations. For notational simplicity the above defined probabilities $P_{Xi}(t')$ or $P_{Xi}(t'')$, Xi = Fi, Mi, or Di, i = 1, 2 or nothing, will be denoted by P_{Xi}' or P_{Xi}'' respectively.

Formulation A: Minimize P_M subject to $P_F = \alpha_o$ and $P_R = \beta_o$.
From equation (49)

$$\alpha_o = P_{F1}'' + P_{F2}'[P_{F1}'-P_{F1}''] \tag{53}$$

from which we obtain,

$$P_{F2}' = \frac{\alpha_o - P_{F1}''}{P_{F1}' - P_{F1}''} \tag{54}$$

and

$$P_{F1}' - P_{F1}'' = \frac{\alpha_o - P_{F1}''}{P_{F2}'} \tag{55}$$

Similarly from equation (51),

$$\beta_o = P_o(P_{F1}' - P_{F1}'') + (1-P_o)[(P_{F1}')^{1/(1+\varepsilon_1)} - (P_{F1}'')^{1/(1+\varepsilon_1)}] \tag{56}$$

Using (55) in (56), we get

$$P'_{F1} = \left[\frac{1}{(1-P_0)} \left[\beta_0 - \frac{P_0(\alpha_0 - P''_{F1})}{P'_{F2}} \right] + (P''_{F1})^{1/(1+\varepsilon_1)} \right]^{(1+\varepsilon_1)} \tag{57}$$

Equations (54) and (56) suggest that minimization of P_M can numerically be achieved using a search algorithm in which we determine the minimum P_M by searching over P''_{F1} in the range of (0,1) and using (56) to determine P'_{F1} subject to the constraint $P'_{F1} \geq P''_{F1}$ (since $t''_1 \geq t'_1$), and (54) to determine P'_{F2}.

The following remarks can be made from the numerical results (see Figures 4.1(a) through 4.1(b)) obtained from this formulation in slowly fading Rayleigh channel:

(a) For fixed probability of false alarm and probability of request, the probability of detection increases monotonically with sensors SNR, Fig. 4.1 (a) and (b).

(b) The P_D curves level off above a certain SNR that depends on the probability of request. From Figure 4.1(a) it is seen that, for higher values of SNR, the performance of the scheme with lower request probability ($\beta_0 = 0.01$), exceeds the performance of the scheme with higher request rate ($\beta_0 = 0.1$). This behavior can be explained by looking at the definition of the indecision region. The primary sensor is forced to maintain an indecision region (Fig. 2) of fixed probability about which the only information relayed to the consultant sensor is ignorance. However, for large SNR's, the primary sensor transmits ignorance, although it may very well be very confident about its 0 or 1 decision when the observations fall in the indecision region. Thus, the information available to the two sensors is tempered in order to satisfy the probability of request requirement, causing the saturation phenomenon.

(c) Formulation A is relatively sensitive to the priors, Fig. 4.2 (a). For low values of β_0, the probability of detection increases monotonically with the prior probability of H_0. However, for higher values of the request probability, e.g. $\beta_0 = 0.1$, P_D decreases slightly with P_0. This behavior suggests that in order to maintain satisfactory performance of the dual-sensor in case that the priors are not known precisely, the lowest possible value of P_0 must be chosen to design the system operating points for low consultation rates. However, the highest P_0 must be chosen in case of high consultation rates. In this way, robustness in the performance of the system to unknown fluctuations in the priors is achieved.

Formulation B : Minimize $P_M + kP_R$ subject to $P_F = \alpha_0$ where k is a constant risk factor.

From equation (49),

$$\alpha_0 = P''_{F1} + P'_{F2}(P'_{F1} - P''_{F1}) \tag{58}$$

or,

$$P'_{F1} = \frac{\alpha_0 + [P'_{F2} - 1] P''_{F1}}{P'_{F2}} \tag{59}$$

In this case the minimization of the objective is achieved by choosing P'_{F2} and P''_{F1} and then calculating P'_{F1} using (60) provided that $P'_{F1} \geq P''_{F1}$. We then calculate P_R using equation (51) and P_M from equation (49). Using the search algorithm, we determine the minimum $(P_M + kP_R)$ by searching over P'_{F2} and P''_{F1} in the range (0, .5) x (0, .5).

The following remarks can be made from the numerical results obtained from this formulation in slowly fading Rayleigh channel; Fig.'s 5.1 (a) through 5.1 (f).

(a) The probability of detection increases monotonically with the SNR , Fig. 5.1 (a) .

(b) Even when consultation is risk free (k = 0), there is a maximum rate of consultation beyond which no benefit can be obtained. According to definition of the indecision region, when the observations fall in the indecision region, S1 declares ignorance. Hence, even if k = 0, the probability of request cannot be one, for it will imply that S1 transmits ignorance even when the true hypothesis is known with probability one (If $P_R = 1$, the indecision region covers the entire decision space).

(c) The request probability decreases monotonically (linearly beyond a certain SNR) with the SNR, Fig. 5.1 (b).

(d) Consultation beyond a certain cost, k = 1.4 in the case of 10^{-5} and 10^{-6} false alarm probabilities, is not beneficial, Figures 5.1 (c) and (d).

(e) The performance of this scheme w.r.t. the priors depends on the consultation cost. For low risk consultation, the consultation rate is high, thus the scheme performance depends critically on the prior probabilities. (See Eq.'s (41), (42) and (43).) Hence P_D should decrease w.r.t P_0. On the other hand, if consultation is expensive, the consultation rate is low, thus the dependence of the scheme performance to the priors relatively small. Hence, the P_D exhibits a region of indifference to fluctuations in P_0, beyond which it rises monotonically with P_0, Fig. 5.2 (a). The probability of request exhibits a similar behavior as a function of P_0, Fig 5.2 (b). From the two figures, it is worth noticing the behavior of scheme B, when k = 0. Under risk-free consultation, the scheme maintains a fixed P_D by increasing P_R linearly w.r.t P_0, up to a certain maximum level, for the same reason that is described (b).

Formulation C : Minimize $P_M + P_F + kP_R$

In order to determine P_M, P_F and P_R we use (48), (49) and (51) respectively. Then we use the search algorithm to determine minimum of

$(P_M + P_F + kP_R)$ over P'_{F1}, P''_{F1} and P'_{F2}.

The following remarks can be made from the numerical results (Figures 6.1.(a) through 6.1.(c)) obtained from this formulation in a slowly fading Rayleigh channel:

(a) The performance of this scheme increases monotonically with the SNR, Fig. 6.1.(a).

(b) Even when consultation is risk-free (k = 0), there is a maximum rate of consultation beyond which no benefit can be obtained.

(c) The consultation cost becomes prohibitive when the risk factor k is equal to or exceeds 1.0.

(d) The appearance of the sum $(P_F + P_M)$ in the objective function does not allow to attain low values of P_F and P_M simultaneously.

(e) From Fig. 6.2.(a), it is seen that the performance of this scheme decreases monotonically with the prior P_0 for all risk factors k between 0 and 1. Hence, this formulation does not provide robustness w.r.t. variations in P_0. Furthermore, the probability of false alarm decreases with P_0. However, due to the term $P_F + P_M$ in the objective function, no improvement in feasible beyond a certein P_F, Fig. 6.2.(b). The dependence of the request probability on P_0 is shown in Fig. 6.2.(c).

V. COMPARISON OF NUMERICAL RESULTSFOR RAYLEIGH CHANNEL

Comparing the performance of the random-request scheme, Fig.'s 3.1 (a) and (b), with the performance of Formulation A scheme (fixed probability of request), Fig.'s 4.1 (a) and (b), it is seen that the latter is far superior to the first. However, the probability of detection of Scheme A depends on the prior probabilities, Fig. 4.2 (a). It is interesting to notice that, for fixed SNR and fixed probability of request β_0, the probability of detection increases up to a certain value of the prior probability that depends on β_0, then drops slightly and finally levels off to some, almost, constant value. However, for all values of the prior probability, the performance of Scheme A is far superior to that of random request. Hence, even if the priors are not known, it is better to operate the dual-sensor configuration according to Scheme A rather than using random consultation. Moreover, the larger the value of β_0, the less sensitive the performance of Scheme A is.

Formulation B yields performance which is comparable to that of Formulation A. Moreover, it is sensitive to the associated communication cost and has the advantage that adjusts the overall decision-making cost according to the siganl strength.

Finally, Formualtion C is not directly comparable with any of the other schemes. A crucial point that seems to be a drawback of this formulation it does not yield low values of P_F and P_M simultaneously.

CONCLUSIONS

The following concluding remarks can be made:

(a) From Formulation A, the best consultation strategy for any SNR is to operate the system just at the onset of saturation with P''_{F1} = ($\triangleq$ PFII1) =α_0. Operating at a higher consultation rate than that given at this point gives very little improvement in system performance. Furthermore, consultation under high SNR is beneficial only if much lower false alarm rates are required.

(b) From Fig. 5.1(d) it appears that the optimum request rate is concave with respect to SNR. Hence, for each value of k, there is a critical SNR above and below which optimum consultation rate decreases. Thus, consultation does not seem to improve the system performance significantly when the signal power assumes extreme values.

(c) Comparision of Formulations A and B shows that for any value of k, the optimum consultation rate depends on constraints and other costs used to define the formulation.

REFERENCES

[1] Tenney, R. R. and Sandell, N. R., Jr., "Detection with Distributed Sensors," IEEE Trans. on Aerospace and Electronic Systems, Vol. AES-17, July 1981, pp. 501-510.

[2] Thomopoulos, S. C. A., Viswanathan, R. and Bougoulias, D. K., "Optimal Decision Fusion in Multiple Sensor Systems," Proceedings of the 24th Allerton Conference, October 1-3, 1986, Allerton House, Monticello, Illinois, pp. 984-993.

[3] Thomopoulos, S. C. A., Viswanathan, R. and Bougoulias, D. K., "Optimal Decision Fusion in Multiple Sensor Systems," IEEE Trans. on Aerospace and Electronic Systems, Sept. 1987, pp. 644-653.

[4] Srinivasan, R., "Distributed Radar Detection Theory," IEE Proceedings, Vol. 133, Pt. F, No. 1, February 1986, pp. 55-60.

[5] Thomopoulos, S. C. A., Viswanathan, R. and Bougoulias, D. K., "Optimal and Suboptimal Distributed Decision Fusion," to be presented at the CDC'87, Los Angeles, Dec. 9-11, 1987. Also submitted to IEEE Trans. on AES.

[6] Viswanathan , R., Thomopoulos, S. C. A. and Tumuluri, R., "Optimal Serial Distributed Decision Fusion," to be presented at the CDC'87, Los Angeles, Dec. 9-11, 1987. Also submitted to IEEE Trans. on AES.

[7] Papastavrou, J. D. and Athans, M., "A Distributed Hypothesis-Testing Team Decision with Communications Cost," Proceedings of 25th Conference on Decision and Control, Athens, Greece, December 1986.

[8] Van Trees, H. L., "Detection, Estimation and Modulation Theory, Part I, " John Wiley & Sons, N.Y., 1968.

[9] DiFranco, J.V., and Rubin, W.L., "Radar Detection," Artech House, MA, 1980.

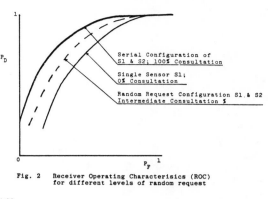

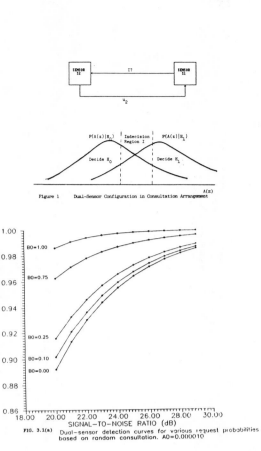

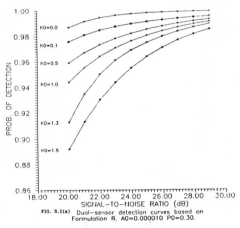

Figure 1   Dual-Sensor Configuration in Consultation Arrangement

FIG. 3.1(a)   Dual-sensor detection curves for various request probabilities based on random consultation. A0=0.000010

Fig. 2   Receiver Operating Characterisics (ROC) for different levels of random request

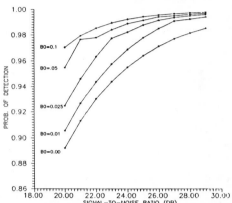

FIG. 4.1(a)   Dual-sensor detection curves for various request probabilities based on Formulation A. A0=0.000010. P=0.3

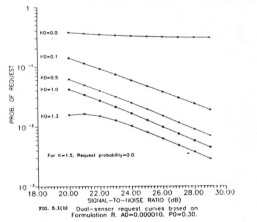

FIG. 5.1(b)   Dual-sensor request curves based on Formulation B. A0=0.000010. P0=0.30.

FIG. 5.1(b)   Dual-sensor detection curves based on Formulation B. A0=0.000010 P0=0.30.

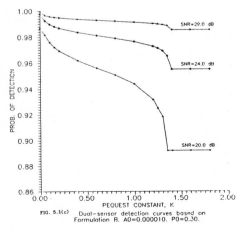

FIG. 5.1(c)   Dual-sensor detection curves based on Formulation B. A0=0.000010. P0=0.30.

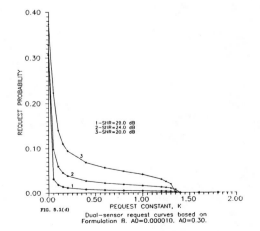

FIG. 5.1(d)   Dual-sensor request curves based on Formulation B. A0=0.000010. A0=0.30.

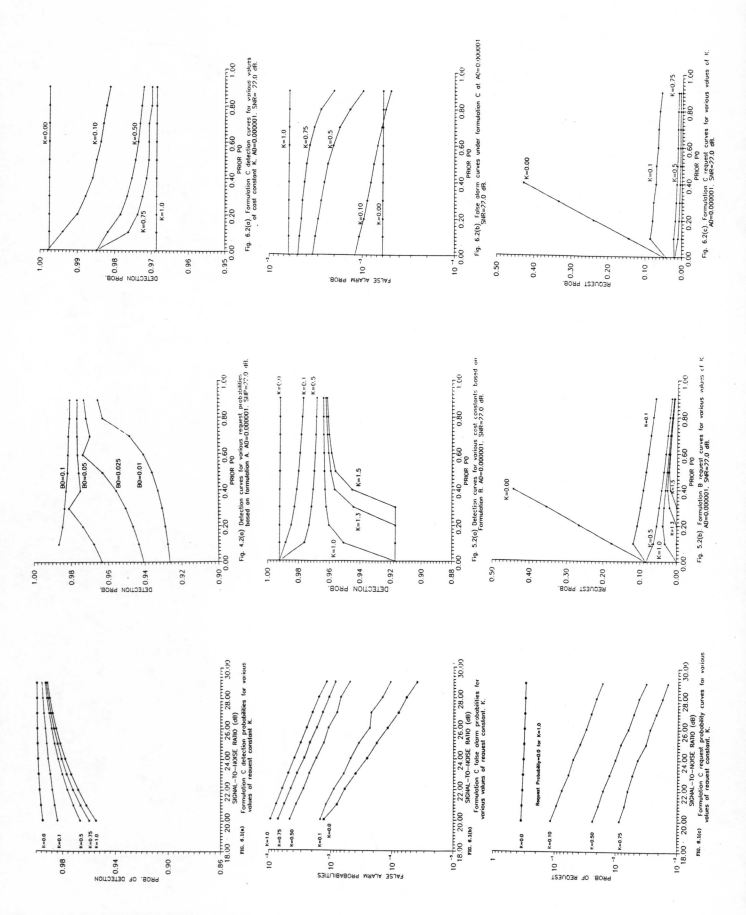

Fig. 6.1(a) Formulation C detection probabilities for various values of request constant K.

Fig. 4.2(a) Detection curves for various request probabilities based on formulation A. A0=0.000001. SNR=77.0 dB.

Fig. 6.2(a) Formulation C detection curves for various values of cost constant K. A0=0.000001. SNR= 77.0 dB.

Fig. 6.1(b) Formulation C false alarm probabilities for various values of request constant. K.

Fig. 5.2(a) Detection curves for various cost constants based on formulation B. A0=0.000001. SNR=77.0 dB.

Fig. 6.2(b) False alarm curves under formulation C at A0=0.000001 SNR=77.0 dB.

Fig. 6.1(c) Formulation C request probability curves for various values of request constant. K.

Fig. 5.2(b) Formulation B request curves for various values of K. A0=0.000001. SNR=77.0 dB.

Fig. 6.2(c) Formulation C request curves for various values of K. A0=0.000001. SNR=77.0 dB.

SENSOR FUSION

Volume 931

Session 1 (continued)

Sensor Fusion Concepts

Chair
Elizabeth Beggs
MAC Associates, Inc.

Multisensor integration and fusion: issues and approaches

Ren C. Luo and Michael G. Kay

Robotics and Intelligent Systems Laboratory
Department of Electrical and Computer Engineering
North Carolina State University
Raleigh, North Carolina 27695-7911

ABSTRACT

Issues concerning the effective integration of multiple sensors into the operation of intelligent systems are presented, and a description of some of the general paradigms and methodologies that address this problem is given. Multisensor integration, and the related notion of multisensor fusion, are defined and distinguished. The potential advantages and problems resulting from the integration of information from multiple sensors are discussed.

1. INTRODUCTION

In recent years there has been a growing interest in the use of multiple sensors to increase the capabilities of intelligent machines and systems. In order for these systems to effectively use multiple sensors there must be some method of integrating the information provided by these sensors into the operation of the system. Typical of the applications that can benefit from the use of multiple sensors are robots and robotic systems, manufacturing automation, autonomous vehicle navigation, automatic target recognition, and other military applications. A recent workshop has focused on the multisensor integration issues involved in manufacturing automation.[1] Common among all of these applications is the requirement that the system intelligently interact with and operate in an unstructured environment without the complete control of a human operator.

Two of the major abilities that a human operator brings to the task of controlling a system are the use of a flexible body of knowledge and the ability to effectively integrate information of different modality obtained through his or her senses. The increasing use of knowledge-based expert systems is an attempt to capture some aspects of this first ability; current research in multisensor integration is an attempt to capture, and possibly extend to additional modalities, aspects of this second ability. Current research in neural networks[2,3] is providing a common paradigm for the interchange of ideas between neuroscience and computer science. Although currently there has not been much published research on the use of neural networks specifically for multisensor integration, Pellionisz[4] has introduced the term "neurobotics" to describe the possible use of brain-like control and representation in robotic systems.

A well-known example of human multisensory integration is ventriloquism, in which the voice of the ventriloquist seems to an observer to come from the dummy. The ability of visual data (the movement of the dummy's lips) to dominate the auditory data coming from the ventriloquist demonstrates the existence of some process of integration whereby information from one modality (audition) is interpreted solely in terms of information from another modality (vision). Howard[5] has reported research that found the discordance between visual and auditory information becomes noticeable only after the source of each has been separated beyond 30° relative to the observer. Not withstanding ventriloquism, the use of information from these two modalities increases the probability of detecting an event in the environment when compared to the use of either modality alone.[5]

Although in humans the processes of multisensory integration have not yet been found, research on the less complex nervous system of the pit viper has identified neurons in this snake's optic tectum (a midbrain structure found in vertebrates) that are responsive to both visual and infrared information.[6] Infrared information from the pit organ, together with visual information from the eye, are represented on the surface of the optic tectum in a similar spatial orientation so that each region of the optic tectum receives information from the same region of the environment. This allows certain "multimodal" neurons to respond to different combinations of visual and infrared information. Certain "or" neurons respond to information from either modality and could be used by the snake to detect the presence of prey in dim lighting conditions, while certain "and" neurons, which only respond to information from both modalities, could be used to recognize the difference between a warm-blooded mouse and a cool-skinned frog. The "and" neurons have been whimsically described as mouse detectors. In evolutionary terms it seems likely that similar integration processes take place in the tectums of most other vertebrates, although at present only Newman and

Hartline's work on pit vipers[6] has been reported.

2. ISSUES IN MULTISENSOR INTEGRATION AND FUSION

2.1. Multisensor integration versus fusion

Multisensor integration, as defined in this paper, refers to the systematic use of the information provided by multiple sensory devices to assist in the accomplishment of a task by a system. An additional distinction is made between multisensor integration and the more restricted notion of multisensor fusion. Multisensor fusion, as defined in this paper, refers to any stage in the integration process where there is an actual combination (or fusion) of different sources of sensory information into one representational format. (This definition would also apply to the fusion of information from a single sensory device acquired over an extended time period.) Although the distinction of fusion from integration is not standard in the literature, it serves to separate the more general issues involved in the integration of multiple sensory devices at the system architecture and control level, from the more specific issues involving the actual fusion of sensory information.

The structure of the network in Figure 1 is meant to represent a general pattern of multisensor integration and fusion. While the fusion of information takes place at the nodes in the figure, the entire structure is an instance of multisensor integration. In the figure, n sensors are integrated into a structure to provide information to the system. The outputs x_1 and x_2 from the first two sensors are fused at the lower left-hand node into a new representation $x_{1,2}$. The output x_3 from the third sensor could then be fused with $x_{1,2}$ at the next node, resulting in the representation $x_{1,2,3}$, which might then be fused at nodes higher in the structure. In a similar manner, the output from all n sensors could be integrated into the structure. The dashed arcs from the system to each node serve to illustrate the possibility of using some type of interactive information from the system (e.g., feedback from a world model) as part of the fusion process. Shown along the right-hand side of the figure is a scale indicating the level of representation of the information at the corresponding level in the structure. Common among most multisensor integration structures is the transformation from lower to higher levels of representation as the information moves up through the structure. At the lowest level raw sensory data enters the sensors and is transformed into information in the form of a signal. As a result of a series of fusion steps, the signal is transformed into progressively more abstract numeric or symbolic representations. This "signals-to-symbols" paradigm is common in computational vision.[7]

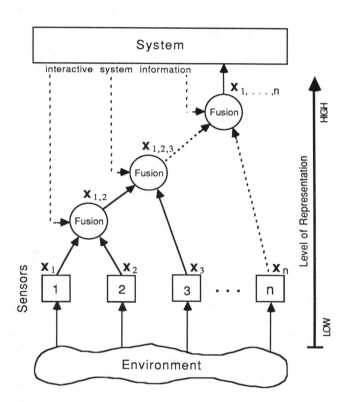

Figure 1. A general pattern of multisensor integration.

2.2. Potential advantages in integrating multiple sensors

The purpose of internal- and external-state sensors is to provide a system with useful information concerning some feature or world model of the system's environment. The potential advantages in integrating information from multiple sensors of disparate sources are that, as a result of the integration, a) information concerning a feature or group of dependent features in the environment can be obtained more accurately, with greater precision, and with more reliability; b) information concerning additional independent features can be obtained; c) information can be obtained at a lesser cost; and d) information can be obtained in less time. These advantages correspond, respectively, to the notions of the redundancy, complementarity, cost, and timeliness of the information. Redundant information, from a group of sensors (or a single sensor over time) concerning either a single feature or group of dependant features in the environment, can serve to increase the accuracy of the fused information by reducing the overall uncertainty of the information. Multiple sensors providing redundant information can also serve to increase the reliability of the system in the case of sensor error or failure. Complementary information from multiple sensors allows independent features in the environment to be used by the system. If the range of possible values a feature can assume is considered as a dimension in a space of features, then the dimension corresponding to each independent feature would be orthogonal to the dimensions of the other independent features in the space. <u>Less costly</u> information, in the context of a system with multiple sensors, is information obtained at a lesser cost when compared to the equivalent information obtained from a single sensor system. Unless the information provided by a sensor is being used for additional functions in the system, the total cost of the single sensor should be compared to the total cost of the integrated multiple sensor structure. <u>More timely</u> information, as compared to the speed at which it could be provided by a single sensor, may result from both the actual speed of operation of each sensor in a group of multiple sensors, or, the possible parallelism involved in the integration process. In practice, some or all of these different advantages will be, to some degree, achievable in a system using multiple sensors.

Figure 2 illustrates the distinction between complementary and redundant information by using the structure from Figure 1 to perform, hypothetically, the task of object discrimination. Four objects are shown in Figure 2a. They are distinguished by the two independent features "shape" and "temperature". Sensors 1 and 2 provide information concerning the shape of an object and sensor 3 provides information concerning its temperature. Figures 2b and 2c show frequency distributions for both square and round objects representing each sensor's historical (i.e., tested) responses to such objects. The "average" square or round object would correspond to the center of these distributions. The bottom axes of both figures represent the range of possible sensor readings. The numerical output values x_1 and x_2 correspond to the "degree of squareness or roundness" of the object, as determined by each sensor. Because sensors 1 and 2 are not able to detect the temperature of an object, objects A and C can not be distinguished and objects B and D can not be distinguished. The dark portion of the axis in each figure corresponds to the range of output values where there is uncertainty as to the shape of the object being detected. The dashed line in each figure corresponds to the point at which, depending on the output value, objects can be distinguished in terms of a feature. Figure 2d is the frequency distribution resulting from the fusion of x_1 and x_2. Without specifying a particular method of fusion, it is usually true that the distribution corresponding to the fusion of dependent information would have less dispersion than its component distributions. The uncertainty in Figure 2d is shown as approximately half that of Figures 2b and 2c. In Figure 2e complementary information from sensor 3 concerning the independent feature temperature is fused with the shape information from Figure 2d (see also Figure 1). As a result of the fusion of information concerning

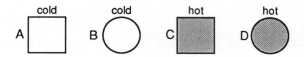

(a) Four objects distinguished by features "shape" and "temperature".

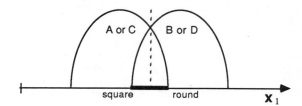

(b) Sensor 1 (shape).

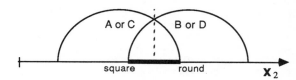

(c) Redundant sensor 2 (shape).

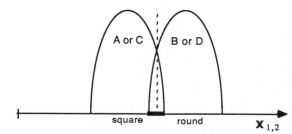

(d) Fusion of sensor 1 and redundant sensor 2.

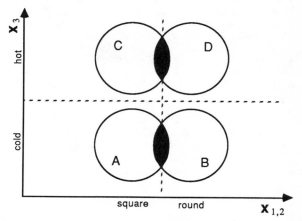

(e) Fusion of sensors 1 & 2 (shape) and complementary sensor 3 (temp.).

Figure 2. Object discrimination using redundant and complementary sensors. (The frequency distribution is two-dimensional in (b), (c) and (d), and three-dimensional in (e).)

this additional feature, it is now possible to discriminate between all four objects. This increase in discrimination ability is one of the advantages resulting from the fusion of complementary information. As indicated in Figure 1, it is also true that the information resulting from this second fusion is at a higher representational level. The result of the first fusion, $x_{1,2}$, is still a numerical value, while the result of the second, $x_{1,2,3}$, could be just a symbol representing one of the four possible objects.

2.3. Potential problems in integrating multiple sensors

Many of the problems associated with creating a general methodology for multisensor integration and fusion, and developing actual systems that use integration, center around methods for handling uncertainty and error. Most of the methods of inference used in fusion require that the uncertainty be commensurate and independent. The uncertainty in the information from different sensors, if it is known at all, may not be independent or commensurate; requiring any method of multisensor fusion to either 1) specify some transformation of the information from each sensor so that it is (approximately) independent and commensurate, or 2) make simplifying assumptions concerning the nature of the uncertainties. A common assumption is that all of the uncertainty is caused by white noise.

Further problems can result from: a) errors during sensor operation (e.g., calibration), b) sensor failure, c) coupling among the components of the integrated system, and d) the control necessary to coordinate the operation of different sensors. The problem of operational control is especially evident during the recognition of remote dynamic objects. All of the sensors have to be controlled so that they remain focused on the object. A typical solution to this problem is to use some type of model of the environment (i.e., a world model). This allows information from the different sensors to be stored in the model at a location corresponding to the source of the information in the environment. Octrees[8] and k-d trees[9] are two of many different types of data structures used to represent world models.

3. APPROACHES TO MULTISENSOR INTEGRATION AND FUSION

3.1. Paradigms and specifications for integration

3.1.1. Hierarchial Phase-Template Approach.
Luo and Lin[10,11] have proposed a general paradigm for multisensor integration in robotic systems based upon four distinct temporal phases in the sensory information acquisition process. The four phases, "Far Away", "Near To", "Touching", and "Manipulation", are distinguished at each phase by the range over which sensing will take place, the subset of sensors typically required, and, most importantly, the type of information desired. During the first phase, "Far Away", only global information concerning the environment is obtained. Typical information at this stage would be the detection, location, or identity of objects in a scene. The most likely types of sensors to be used during this phase would be non-contact sensors like vision cameras and range finding devices. If the scene is found to be of sufficient interest during the first phase, the manipulator can zoom in to obtain more detailed information. This corresponds to the second phase, "Near To". Usually at this close range it is not possible to see the entire object, so non-contact sensors like proximity sensors or "eye-in-hand" vision systems, mounted on the gripper of the manipulator, are used. If it is desired to confirm or integrate the information from the previous two phases, one can proceed to the third phase, "Touching". Contact sensors such as tactile sensors would typically be used at this phase. Finally, if it is necessary to manipulate the object, one can proceed to the fourth phase, "Manipulating". Sensors providing information concerning force/torque, slippage, and weight would typically be used during manipulation.

The information acquired at each phase is represented in the form of a distinct frame-like template. Each template represents information that is both common to all phases (e.g., position and orientation of an object) and specific to the particular phase. During each phase of operation, the information acquired by each sensor is stored as an instance of that phase's template. The information from each sensor can then be fused into a single instance for the phase (see section 3.3.2., below).

3.1.2. Logical sensors.
A "logical sensor", as proposed by Henderson and Shilcrat[12] and then extended[13], is a specification for the abstract definition of a sensor. Through the use of an abstract definition of a sensor the unnecessary details of the actual physical sensor are separated from their functional use in a system. In a manner similar to how an abstract data type separates from the user the unnecessary algorithmic details of operations on data, or, how an operating system separates the peculiarities of a specific printer through the use of a device driver, the use of logical sensors can provide any multiple sensor system with both portability and the ability to adapt to technological changes in a manner transparent to the system.

3.1.3. The NBS sensory and control hierarchy. The Center for Manufacturing Engineering at the National Bureau of Standards (NBS) is implementing an experimental factory called the Automated Manufacturing Research Facility (AMRF). As part of the AMRF, a multisensor interactive hierarchial robot control system is being developed.[14,15,16] As shown in Figure 3, the structure of the control system in AMRF consists of an ascending sensory processing hierarchy (S) coupled to a descending control processing hierarchy (C) via world models (W) at each level. The amount of processing at each level is reduced by the use of a priori knowledge from a world model. A world model can provide predictions to the sensory system concerning incoming information or data that can reduce the amount of processing required. A world model also provides the control system with immediate information so that it does not have to wait on a particular sensory analysis to finish. Use of a world model also promotes modularity because the specific informational requirements of the sensory and control systems are decoupled.

In Figure 3, the hierarchy operates by having raw sensory data enter the system at S_1. At this lowest level, most of the required sensory processing will be continuous monitoring. Any deviation between the actual and expected data at S_1 is sent as feedback information to C_1 and as summary information to the next level S_2, where it again interacts with world model expectations. At the very highest level in the system, an a priori world model and an overall control goal serve to filter down to lower levels both expected and desired information values. It is at the intermediate levels where both of these information flows meet and interact. Based upon current sensory information, the world model is updated. This updated world model can then serve to modify the desired control action until, at the lowest level (C_1), the necessary information is sent to initiate actions in the environment.

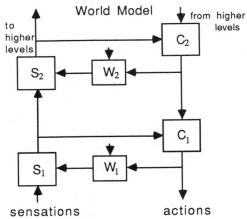

Figure 3. The NBS sensory and control hierarchy (adapted from Kent and Albus[15], and Albus[16]).

3.2. General methods for integration

3.2.1. Adaptive learning. Miller, Glanz, and Kraft[17,18] have applied the adaptive learning approach to the multisensor-based control of robotic manipulators. In experiments using this approach the performance accuracy was limited by the resolution of the sensor feedback, rather than by any limitations in the control structure. Adaptive learning is a method of control in which the system "discovers" the appropriate signals for control based upon the output of the sensors. The system is taught a representative sample of correlated control signals and associated sensory outputs over the range of signals and sensory outputs encountered by the system. Based upon the associations developed during this teaching phase, it is possible to have the system respond to any combination of sensory outputs with the appropriate associated control signal. The system requires no a priori knowledge of the relationship between the structural kinematics of the robot or the desired control signals, and their associated sensory outputs. It is this feature of the adaptive learning approach that makes it attractive when there are possibly multiple sensors interacting to produce complex output.

3.2.2. Object-oriented programming. Allen[19] has used an object-oriented programming methodology to develop a uniform framework for implementing multisensor robotic tasks. Objects are represented by "classes", and "methods" are used to invoke specialized sensor processing procedures based upon the sensor's attributes and behavior. This allows possibly different physical sensors, that have the same behavior, to be used interchangeably. Rodger and Browse[20] have also used the object-oriented approach for multisensory robotic object recognition.

3.2.3. Distributed blackboard. Harmon, Bianchini, and Pinz[21] have used a blackboard architecture to allow economical communication between the distributed sensory subsystems in their integrated multisensor system. Each subsystem sends time-stamped summary output to a blackboard where it becomes available to any fusion process, as well as other subsystems. The time-stamp on the output in the blackboard allows for the fusion of information from fast-changing situations. The blackboard can also contain any of the system information needed by the subsystems. Any number of different fusion methods can be implemented using the output from the blackboard. Harmon, et al.,[21] have used their system to compare different methods of multisensor fusion.

3.3. General methods for fusion

The methods presented in this section are distinguished by being general methods for multisensor fusion. A large number of fusion techniques are available that apply only to a specific group of different sensors, e.g., vision and touch.[22] The sequence in which the methods described in this section are presented correspond to their use of increasing higher levels of representation (see Figure 1). The representations used extend from low-level probability distributions for statistical inference to high-level production rules for logical inference.

3.3.1. Statistical pattern recognition.

Pau[23] describes a number of techniques that reduce the error in classifying objects by using multiple sensors to provide complementary information concerning independent features. To avoid an exponential increase in complexity as sensors are added to a system, a key requirement is that the number of features and levels in the recognition process increase at a slower rate than the number of sensors. To meet this requirement it becomes necessary to improve the overall methods of feature extraction and selection; thus, multisensor fusion becomes a problem in the context of statistical pattern recognition. Pau describes a number of operators and techniques that can merge the features from different sensors to limit the growth in the number of features.

3.3.2. Bayesian estimation using consensus sensors.

Luo and Lin[10,11,24] have developed a method for the fusion of redundant information from multiple sensors that can be used within their hierarchial phase-template paradigm (see section 3.1.1., above). The central idea behind the method is to first eliminate from consideration the sensor information that is likely to be in error and then use the information from the remaining consensus sensors to calculate a Bayesian estimator. Given certain assumptions, this estimator will provide the optimum estimation of the fused information.

Figure 4 shows a functional block diagram of the method. The information from each sensor is represented as a probability density function. Given readings from n sensors in the system, the resulting information is first made commensurate by preprocessing. An n by n _distance matrix_ is created by calculating for each element (i, j) in the matrix the _confidence distance measure_ between the information from sensors i and j. If the density functions are assumed to be Gaussian this distance can be computed by the use of the error function. Empirically determined threshold values are then applied to the elements in the matrix. Elements not exceeding their threshold are represented by a 1 in a binary-valued n by n _relation matrix_. The largest connected di-graph formed from this matrix will determine the group of _consensus sensors_ most likely not to be in error. The optimal fusion of the information is determined by finding the Bayesian estimator that maximizes the likelihood function of the consensus sensors.

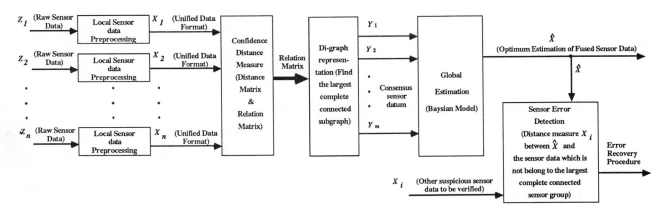

Figure 4. Functional block diagram of the consensus sensor fusion method.

3.3.3. Multi-Bayesian.

Durrant-Whyte[25] has developed a model of a multisensor system that represents the task environment as a collection of uncertain geometric objects. Each sensor in the system is described by its ability to extract useful descriptions of these objects. A contaminated Gaussian distribution is used to represent the geometric objects, which, after removing a certain outlying fraction of the distribution, approximates a Gaussian distribution. The sensors in the system are considered as a team of decision makers. Together, the sensors must determine a team-consensus view of the environment. A multi-Bayesian approach, with each sensor considered as a Bayesian estimator, is used to combine the associated probability distributions of each respective object into a joint posterior distribution function. A likelihood function of this joint distribution is then maximized to provide the final fusion of the sensory information. The fused information, together with an a priori model of the environment, can then be used to direct the robotic system during the execution of different tasks.

3.3.4. Shafer-Dempster evidential reasoning.

Garvey, Lowrance, and Fischler[26] introduced the possibility of using Shafer-Dempster evidential reasoning in multisensor fusion. Bogler,[27] and Waltz and Buede[28] have explored its possible application in, respectively, multisensor target identification, and military command and control. Shafer-Dempster evidential reasoning[29] is an extension to the Bayesian approach that makes explicit any lack of information concerning a proposition's probability by separating firm support for the proposition from just its plausibility. In the Bayesian approach, all propositions (e.g., features in the environment) for which there is no information are assigned equal a priori probability. When additional information from a sensor becomes available and the number of unknown propositions is large relative to the number of known propositions, an intuitively unsatisfying result of the Bayesian approach is that the probabilities of known propositions become unstable. In the Shafer-Dempster approach this is avoided by not assigning unknown propositions an a priori probability; instead, these propositions are assigned to "ignorance". Probabilities are assigned to these propositions, thus reducing ignorance, only when supporting information becomes available.

3.3.5. Fuzzy logic.

Huntsberger and Jayaramamurthy[30] have used fuzzy logic to fuse information for scene analysis and object recognition. Fuzzy logic,[31] a type of multiple-valued logic, allows the uncertainty in multisensor fusion to be directly represented in the inference (i.e., fusion) process by allowing each proposition, as well as the actual implication operator, to be assigned a real number from 0.0 to 1.0 indicating its degree of truth. Consistent logical inference can take place if the uncertainty of the fusion process is modeled in some systematic fashion.

3.3.6. Production rules.

Kamat,[32] and Belknap, Riseman, and Hanson[33] have used rule-based systems for object recognition using multisensor fusion. Production rules, with an associated confidence factor to indicate a degree of uncertainty, are used to symbolically represent the relation between an object feature and the corresponding sensory information. Fusion takes place when two or more rules, referring to the same object, are combined during logical inference to form one rule. The major problem in using rule-based methods for fusion is that the confidence factor of each rule is dependent upon the other rules in the system, making it difficult to alter the system when conditions change.

4. CONCLUSION

The issues of multisensor integration and fusion discussed in this paper were stated in the broadest possible terms in the hope that many of the different approaches to this problem could be presented and compared using a common terminology. The approaches presented were selected because each is a general approach, and, together, they demonstrate the wide scope of possible methods and techniques that can be effectively used to address the problem of integrating multiple sensors.

5. REFERENCES

1. T.C. Henderson, P.K. Allen, A. Mitiche, H. Durrant-Whyte, and W. Snyder, eds., Workshop on multisensor integration in manufacturing automation, Tech. Rep. UUCS-87-006, Univ. of Utah, Dept. of Computer Sci., Snowbird, Utah, February (1987).
2. G.E. Hinton and J.A. Anderson, eds., Parallel models of associative memory, Erlbaum, Hillsdale, NJ (1981).
3. D.E. Rumelhart and J.L. McClelland, eds., Parallel distributed processing: explorations into the microstructure of cognition, 2 vols., MIT Press, Cambridge, Mass. (1986).
4. A.J. Pellionisz, "Sensorimotor operation: a ground for the co-evolution of brain-theory with neurobotics and neurocomputers", in Proc. 1st IEEE Intl. Conf. on Neural Nets, 593-600, San Diego, June (1987).
5. I.P. Howard, Human visual orientation, pp. 443-459, Wiley, Chichester (1982).
6. E.A. Newman and P.H. Hartline, "The infrared "vision" of snakes", Scientific American, 246(3), 116-127, March (1982).
7. M.A. Fischler and O. Firschein, Intelligence: the eye, the brain and the computer, pp. 241-242, Addison-Wesley, Reading, Mass. (1987).
8. S. Chen, "Multisensor fusion and navigation of mobile robots", Intl. J. of Intelligent Systems, 2(2), 227-251 (1987).
9. T.C. Henderson, W.S. Fai and C. Hansen, "MKS: a multisensor kernel system", IEEE Trans. On Systems, Man, and Cybernetics, 14(5), 784-791 (1984).
10. R.C. Luo, M. Lin and R.S. Scherp, "The issues and approaches of a robot multi-sensor integration", in Proc. IEEE Intl. Conf. Robotics and Auto., 1941-1946, Raleigh, March (1987).

11. R.C. Luo, M. Lin and R.S. Scherp, "Dynamic multi-sensor data fusion system for intelligent robots", <u>IEEE J. of Robotics and Automation</u>, (in press).

12. T. Henderson and E. Shilcrat, "Logical sensor systems", <u>J. of Robotic Sys.</u>, 1(2), 169-193 (1984).

13. T. Henderson, C. Hansen and B. Bhanu, "A framework for distributed sensing and control", in <u>Proc. 9th Intl. Joint Conf. on Artificial Intelligence</u>, 1106-1109, Los Angeles, August (1985).

14. H.G. McCain, "A hierarchically controlled, sensory interactive robot in the automated manufacturing research facility", in <u>Proc. IEEE Intl. Conf. on Robotics and Automation</u>, 931-939, St. Louis, March (1985).

15. E.W. Kent and J.S. Albus, "Servoed world models as interfaces between robot control systems and sensory data", <u>Robotica</u>, 2, 17-25 (1984).

16. J.S. Albus, <u>Brains, behavior, & robotics</u>, Byte Books, Peterborough, NH (1981).

17. W.T. Miller III, F.H. Glanz and L.G. Kraft III, "Application of a general learning algorithm to the control of robotic manipulators", <u>Intl. J. of Robotics Research</u>, 6(2), 84-98 (1987).

18. W.T. Miller III, "Sensor-based control of robotic manipulators using a general learning algorithm", <u>IEEE J. of Robotics and Automation</u>, 3(2), 157-165 (1987).

19. P.K. Allen, "A framework for implementing multi-sensor robotic tasks", in <u>Proc. ASME Intl. Computers in Engr. Conf. and Exhibition</u>, R. Raghavan and T. J. Cokonis, eds., 303-309, New York, August (1987).

20. J.C. Rodger and R.A. Browse, "An object-based representation for multisensory robotic perception", in <u>Proc. Workshop on Spatial Reasoning and Multi-Sensor Fusion</u>, 13-20, St. Charles, Ill., October (1987).

21. S.Y. Harmon, "Sensor data fusion through a distributed blackboard", in <u>Proc. IEEE Intl. Conf. on Robotics and Automation</u>, 1449-1454, San Francisco, April (1986).

22. P.K. Allen, <u>Robotic object recognition using vision and touch</u>, Kluwer, Boston (1987).

23. L.F. Pau, "Fusion of multisensor data in pattern recognition", in <u>Pattern recognition theory and applications</u>, J. Kittler, K.S. Fu and L.F. Pau, eds., pp. 189-201, Reidel (1982).

24. R.C. Luo and M. Lin, "Robot multi-sensor fusion and integration: optimum estimation of fused sensor data", to appear in <u>Proc. IEEE Intl. Conf. on Robotics and Automation</u>, Philadelphia, April (1988).

25. H.F. Durrant-Whyte, "Consistent integration and propagation of disparate sensor observations", <u>Intl. J. of Robotics Research</u>, 6(3), 3-24 (1987).

26. T.D. Garvey, J.D. Lowrance and M.A. Fischler, "An inference technique for integrating knowledge from disparate sources", in <u>Proc. 7th Intl. Joint Conf. on Artificial Intelligence</u>, 319-325, Vancouver, August (1981).

27. P.L. Bogler, "Shafer-Dempster reasoning with applications to multisensor target identification systems", <u>IEEE Trans. on Systems, Man, and Cybernetics</u>, 17(6), 968-977 (1987).

28. E.L. Waltz and D.M. Buede, "Data fusion and decision support for command and control", <u>IEEE Trans. on Systems, Man, and Cybernetics</u>, 16(6), 865-879 (1986).

29. G. Shafer, <u>A mathematical theory of evidence</u>, Princeton Univ. Press, Princeton, NJ (1976).

30. T.L. Huntsberger and S.N. Jayaramamurthy, "A framework for multi-sensor fusion in the presence of uncertainty", in <u>Proc. Workshop on Spatial Reasoning and Multi-Sensor Fusion</u>, 345-350, St. Charles, Ill., October (1987).

31. L.A. Zadeh, "Fuzzy sets", <u>Information and Control</u>, 8, 338-353 (1965).

32. S.J. Kamat, "Value function structure for multiple sensor integration", in <u>Proc. SPIE Vol. 579 Intell. Robots and Computer Vision</u>, 432-435, Cambridge, Mass., September (1985).

33. R. Belknap, E. Riseman and A. Hanson, "The information fusion problem and rule-based hypotheses applied to complex aggregations of image events", in <u>Proc. IEEE Conf. on Computer Vision and Pattern Recog.</u>, 227-234, Miami Beach, June (1986).

Theoretical approaches to data association and fusion

Samuel S. Blackman

Hughes Aircraft Company
2141 E. Rosecrans Blvd. (Mail Station
E52/C227
El Segundo, California 90245

ABSTRACT

This paper presents a survey of the issues and methods for the association and fusion of data from multiple sensors. It will cover three broad areas. The first, data association, refers to the problem of partitioning the sensor data into tracks according to source. We next discuss the techniques used for kinematic and attribute estimation after data association has been performed. In particular, we will discuss the philosophies associated with alternative techniques for attribute and target type estimation, with an emphasis upon the Dempster-Shafer approach. The third major topic discussed is sensor allocation.

1. INTRODUCTION

This paper presents a survey of the theoretical methods for the association and fusion of data from multiple sensors. There exists a rather well developed theory for the design and analysis of individual sensors. However, the methods required to efficiently utilize the data from multiple sensors for such tasks as tracking and identifying targets are not nearly as well developed. The goal of this paper is to present a broad overview of the many methods that are commonly in use and to discuss the relative merits of these methods for future advanced systems.

There are two basic areas of multiple sensor data fusion. The first area is concerned with the process of track formation and estimation of target quantities, such as position and velocity, and certain attributes, such as engine type. The second area, which typically involves the use of artificial intelligence methods, refers to making higher order inferences, such as target intent, enemy order of battle, etc., given the estimated track quantities. This paper will be primarily concerned with the first area.

An important area in the field of multiple sensor tracking is the allocation of sensors to obtain new data. This important topic has only recently begun to be addressed in a theoretical, rather than an ad hoc manner. This paper will outline the utility theory and expert system approaches to sensor allocation.

2. DATA ASSOCIATION

Data association, with the goal of partitioning observations into tracks, is the key function of any surveillance system. The importance of this subject has led to the recent publication of three books [1,2,3] and numerous papers that extensively discuss the issues. This section will summarize the philosophies and methods in current application.

2.1 Nearest-neighbor versus all-neighbors data association

The nearest-neighbor approach to data association determines a unique pairing so that at most one observation can be paired with a previously established track. Also, a given observation can only be used once, either to update an existing track or to start a new track. This method is based upon likelihood theory and the goal is to minimize an overall distance function that considers all observation-to-track pairings that satisfy a preliminary gating test. However, since for a dense target environment there may be many sets of pairings that are computed to have similar likelihoods, the probability of error is typically large for this approach.

The alternative all-neighbors approach incorporates all observations within a gated region about the predicted track position into the update of that track. Also, a given observation can be used to update multiple tracks. Track update is then based on a probabilistically determined weighted sum of all observations within the gated region.

The all-neighbors approach, in effect, performs an averaging over observation-to-track data association hypotheses that have roughly comparable likelihood. This approach has been shown to be very effective for the tracking of single targets, using single or multiple sensors, against a background of clutter or noise. However, when considering multiple target environments it may be necessary to devise special logic for a particular application so that new target track initiation and target resolution capabilities are maintained.[4]

2.2 Sequential versus deferred decision logic

An ideal, batch processing, approach that would process all observations (from all time) together is typically not computationally feasible. Thus, a more standard approach has been to perform processing in a recursive (or sequential) manner as data are received. For example, a set of observations from a time frame (or scan) of a scanning radar may be collected and observation-to-track data association performed on these observations during the time period when the sensor is collecting the next frame of data. Using this approach, once data association decisions are made, they are irrevocable.

A deferred decision approach to data association, as exemplified by the multiple hypothesis tracking (MHT) method[5], allows the final decision on difficult data association alternatives to be postponed until more data are received. Alternative hypotheses are formed and re-evaluated when later data are received. This approach clearly has the potential for ultimately achieving a much higher correct decision probability than does the sequential decision method. However, in order to maintain computational feasibility, an intricate logic to delete (prune) unlikely hypotheses and to combine similar hypotheses is required.

2.3 Issues in multiple sensor data association

The use of multiple sensor data has the potential, through the inclusion of more varied data, to greatly improve data association. For example, the combined use of accurate range data from a radar and accurate angle information from an optical sensor can lead to accurate track position estimates so that data association uncertainty is reduced. However, the problem of initially combining radar and optical sensor data into a track when the latter does not contain any range information is difficult.

An additional problem with the association of multiple sensor data occurs with sensors of different resolution. For example, an optical sensor may resolve closely spaced targets that appear as a single target to the radar. A similar problem occurs when combining data from a radar and an electronic support measures (ESM) sensor because the platform being tracked may contain multiple emitters that are detected independently by the ESM.

The special problems involved in data association with multiple sensors typically require a complex logic utilizing the deferred decision approach. An example is the logic which has been developed for combining radar and ESM data[6]. Using this method, the ESM observation data are compared with radar track estimates. Then, a set of rules is applied so that three levels of tentative decision are included. In addition to tentative decisions, the firm decisions to pair an ESM track with one of the radar tracks or to determine the ESM track to not have a corresponding radar track are available.

3. LEVEL OF DATA ASSOCIATION: CENTRAL-VERSUS SENSOR-LEVEL TRACKING

A key issue when developing a multiple sensor surveillance system is to decide where data association is to first occur. One alternative is to have each sensor maintain its own track file, based only upon the observations received by that sensor. Eventually these sensor-level tracks must be combined to form a central-level track. The alternative to sensor-level tracking is for all observation data to be sent directly to a central processor where a master track file is formed using all the observation data. For the later approach, no track fusion is required.

There are a number of trade-offs involved in the choice of sensor-versus central-level tracking. The central-level tracking approach is more accurate because all data is processed at the same place resulting in less track uncertainty and a corresponding decrease in misassociation. Also, mean tracking errors for dynamic targets are decreased because of the effective higher data rate. When track fusion occurs for the sensor-level tracking method there is a reduction in the random error but a mean error, for example due to target maneuver, cannot be averaged out.

The major objection to central-level tracking is the high potential data transfer rate. Note that the periodic transmission of tracks, required for sensor-level tracking, typically involves much less data transmission than if all the observations that formed the tracks were transferred. Other potential problems with central-level tracking relate to the military environment where vulnerability to central-level failure (or destruction) and corrupted sensor data are important practical issues.

A major difficulty associated with the use of sensor-level tracking arises in the ultimate combining of the sensor tracks. First, an extensive track-to-track association logic must be defined. This may be difficult if deferred decision data association logic is used, because multiple hypotheses logic must be manipulated at both sensor and combined tracking levels. Finally, even if correct track association is performed, the combining logic must account for common error sources due to potential target maneuvers[7].

Since both the sensor-and central-level approaches suffer from potential problems, a combined approach may be proposed where both central-and sensor-level tracks are maintained. One technique is to initially form central-level tracks from sensor-level tracks and to then send selected observations to the central-level for track update. Another potential saving associated with this method occurs if observations from different sensors, but from the same target, can be precombined before inclusion in the processing.

4. KINEMATIC ESTIMATION

Kinematic filtering and prediction is used

to determine such quantities as target position and velocity. Kinematic filters are required to determine the weighting to be applied to incoming measurement data and to estimate future target position and velocity.

4.1 Kalman versus suboptimal filtering

Kalman filtering provides a convenient means for determining the weightings (denoted as gains) to be given to input measurement data. It also provides an estimate of the target tracking error statistics, through the covariance matrix. The Kalman filter covariance matrix can then be used, along with measurement error statistics, for gating and for determining the final observation-to-track assignment.

The complexity of Kalman filtering has often led to the use of fixed-coefficient filters where some set of "best" filter gains are precomputed and used in place of the Kalman gains. However, since the Kalman gains vary adaptively in accordance with the input measurement rate and accuracy and with the target geometry, the use of fixed gain filters typically leads to highly suboptimal performance.

4.2 Passive (angle-only) sensor tracking

The design of tracking filters for radar data is relatively straight-forward, because the radar, through range and angular measurements, provides a three-dimensional target position. However, passive sensors such as infrared search and track sets (IRST) and radar warning receivers (RWR) only provide angular measurements. Thus, the full three-dimensional target position information may not be available with a single passive sensor.

One approach to tracking with a passive sensor is to merely be content with angle information, and, for example, just track angle and angle rate. Alternatively, it may be possible to estimate a full three-dimensional target state vector if there is an ownship maneuver and if the target does not maneuver.[8] A third alternative is to use multiple angular position estimates from sensors at different locations to calculate target range through triangulation. However, for the case where there are multiple targets in a common plane with the sensors a major problem with triangulation is the occurrence of false intersections or "ghosts".

4.3 Combining active and passive sensor data

In military applications, it is often necessary to restrict active transmission in order to avoid alerting the enemy to one's presence and position. One approach to this problem is to limit active sensor transmission providing sporadic active sensor data to augment the passive data. A typical application of this approach would use a radar to aid in a system where an infrared search and tracking (IRST) system is the primary sensor for track initiation and maintenance. The performance associated with this approach is highly dependent upon the target range and maneuver capability. However, for typical airborne target tracking problems, results have indicated that little angle tracking degradation occurs even when the rate of active sensor data (range and range rate) is as low as one-tenth the rate of the passive angle measurement data.

5. FUSION OF MULTIPLE SENSOR ATTRIBUTE AND TARGET TYPE DATA

Future multi-sensor systems will provide a variety of target attribute data and a major current research area is the development of techniques to combine (or fuse) this data. Typically the ultimate goal of this fusion process is to determine target type and identity.

Pattern recognition is one method of using multiple sensor data to determine target identification. Using this approach, one would determine the appropriate set of features to be formed from multiple sensor observation data and the best weighting (confidence level) to be used with this data. This, however, is a complex process and is probably only feasible for a limited number of sensors. Also a redefinition of the features and weightings will be required when another sensor is added to the system.

An alternative to the pattern recognition approach is for each sensor to first process its own data and to then output its current best estimate of target attributes. Here, the term attribute will include both target characteristics, such as engine type, and the actual target ID type. The confidence associated with each output must also be transmitted or assumed to be known from previous experience. Thus, the problem becomes one of combining multiple sensor attribute data to specify target type and the associated confidence level.

The simplest approaches to combining multiple sensor data are heuristic methods, such as voting and the use of possibility theory. The later approach starts with unity possibility for all possible target types. Then, there is effectively a process of elimination as the possibilities are decreased when sensor data is inconsistent with particular target types. Two alternative mathematical approaches for fusion are the Bayesian[9] and Dempster-Shafer methods[10].

5.1 Bayesian approach

The application of a Bayesian approach to the attribute and target identification problem requires priori information and conditional probabilities. First, the measurement process is defined by $P(X_m/X)$ which is the, assumed known, probability of receiving measurement X_m given that the true quantity is X. Then, whenever measurement data are received, the updated probabilities for quantities X are computed using Bayes' rule.

To summarize, Bayes' rule is solved recursively as new data are received. This relatively simple relationship provides the estimated probabilities of target attribute based directly on the measurements of involved quantities. However, to initiate the process, before any received measurements, requires an initial estimate of the probability of X. Also, the estimated probabilities can be improved using known interrelationships, as expressed by conditional probabilities, between attributes and target types.

Target type estimates can be used to update target attribute estimates and vice versa. This can be accomplished by virtue of the fact that certain types of targets utilize specific types of equipment. For example, the radars used in passenger type aircraft are different than the radars used in military fighters. In order to relate attribute and target type estimation, it is first necessary to specify conditional probability relationships between expected attributes and target types. Then the problem of jointly estimating target type and attribute values can be formulated as a special case of a general inference net[2]

5.2 Dempster-Shafer evidential reasoning

Considerable interest has recently arisen regarding the application of the Dempster-Shafer method to the problem of multiple sensor fusion for the purpose of target identification[10]. This approach has a number of unique features and advantages that are probably best illustrated by example.

Assume that an imaging electro-optical (EO) sensor declares a given target to be of a shape category that is consistent only with targets types T_1 and T_2, which are hostile and friendly, respectively. Further, assume the confidence on this declaration to be 0.9. According to this declaration, there will be a mass $M_1(T_1 V T_2) = 0.9$ assigned to the disjunction of T_1 and T_2 (either T_1 or T_2). Also, a mass $M_1(\Theta) = 0.1$, corresponding to one minus the confidence of 0.9, is assigned to Θ, which represents the disjunction of all propositions (all target types).

Next, assume that an IFF (Interrogation Friend or Foe) sensor does not receive a response from the target in question. This implies a hostile (or at least nonfriendly) target type but, since there are conditions under which a friend may not respond to an IFF interrogation, the confidence on the hypothesis of a hostile target is only 0.6 and the remainder is assigned to Θ, $M_2(\Theta) = 0.4$.

Note that the Dempster-Shafer method has allowed each sensor to contribute information on its own terms, or more formally at its own level of abstraction. Probability mass can be assigned to individual propositions or to the disjunction of propositions. For example, there is no need to arbitrarily force the target type probability mass associated with the shape declaration into equal probabilities for types T_1 and T_2.

Dempster's rule is used to combine sensor information by effectively determining the intersection of the sensor statements. For example, taking the product of $M_1(T_1 V T_2)$ and $M_2(H)$ results in a mass declaration of 0.54 to the proposition that the target type is T_1 (a hostile target). Similarly, the product of $M_1(\Theta)$ and $M_2(H)$ results in a declaration of 0.06 to hostile. The resulting probability mass vector is:

$$(1) \quad \underline{M} = \begin{bmatrix} M(T_1) = 0.54 \\ M(T_1 V T_2) = 0.36 \\ M(H) = 0.06 \\ M(\Theta) = 0.04 \end{bmatrix}$$

Referring to the mass vector given in Eq. (1) illustrates the Dempster-Shafer concept of a probability interval, defined by the support and plausibility for a given proposition. The support for a given proposition is that mass assigned directly to that proposition. For example, the support for the proposition that the target type is T_1 is 0.54. The plausibility of a given proposition is the sum of all masses not assigned to its negation. For example, noting that Θ contains all propositions, the plausibility of the type T_1 proposition is 1.0 On the other hand, the support and plausability for the proposition that the target is the friendly type T_2 are 0.0 and 0.4, respectively.

Dempster's rule also conveniently handles the condition of inconsistency between sensor declarations. For example, if T_1 and T_2 were both friendly, the above example would have assigned the value of 0.54 to inconsistency (denoted k),

$$M_1(T_1 V T_2) \ M_2(H) = 0.54 = k$$

Then, before final mass values are declared a normalization by the factor 1-k is required.[2,10]

6. DISTRIBUTED SENSOR SYSTEMS

The discussion so far has been basically for local sensor fusion which can be defined as the integration of data from sensors on board a single platform. Next, we briefly consider the issues involved in the fusion of data from sensors that are spatially distributed so that they see the targets from different locations and viewing angles.

An important consideration for a distributed sensor system is the communication involved in transfer of information between the sensors. This can lead to a trade-off between allocating system resources for computation versus communications. Also, the potential cost of transferring all observations and the desirability of performing local data processing tends to favor the sensor-level tracking approach. However, techniques have been proposed in which selected observations are passed between sensor sites, so the central-level tracking accuracy can be achieved by distributed data processing.[11]

Another important issue is that of sensor control. Here, based upon survivability considerations, it is desirable that each individual sensor system have the capability to operate autonomously. This implies that each site will determine, based upon locally available information, where its sensors should point and what targets should be tracked.

Another important question involves track maintenance. Should all tracks be maintained at all sites or should each site only maintain a selected subset of target tracks? One common approach to this problem has each site maintaining two sets of track files. The first (global) track file contains tracks on all targets, with track information being communicated from other sites. The second (local) track file contains tracks on those targets that are being actively tracked by that site. Thus, the global track file is a composite of the local track files which gives each site an overview of the entire situation. It aids in local sensor allocation and in the determination of which targets should be included in the local track file.

7. SENSOR ALLOCATION

The simplest approach to sensor allocation uses one sensor to cue another. For a typical scenario, a low angular resolution sensor with a large field of view obtains a preliminary estimate of target position. Then, a second high resolution (angular resolution) sensor is slewed to the predicted target position. This approach may be appropriate if the number of tracks requiring sensor allocation is limited. However, for a dense target environment, more sophisticated approaches are required.

A modern sensor allocation algorithm will be required to adaptively direct multi-sensor systems: (1) to search for new targets, (2) to update target positions, (3) to obtain target identification, and (4) to remain covert and thus evade detection by the enemy of its presence. This algorithm should be adaptive to varying target densities and dynamics, to sensor capabilities, and to the overall situation and mission objectives.

There are two general approaches to sensor allocation in a dense target environment. The first is based upon longer term planning and involves computing the expected frequency at which the many required tasks should be performed. For example, a regular update rate would be computed for each target track. The next step will involve bin or time slot

packing to accommodate long term objectives to be performed by each sensor.

An alternative approach is to perform allocation more dynamically based upon current conditions. For example, rather than adhering to some predetermined track update rate, the desirability of updating each track is regularly computed based upon current conditions and compared with the desirability of performing other tasks, such as search or communication. Then, short term objectives are formed and frequently updated. We will next discuss two techniques available for implementing the dynamic sensor allocation approach.

7.1 Utility theory based allocation

Sensor allocation can be based upon the minimization of some cost function or equivalently the maximization of a utility function. The basic approach here is to determine the usefulness of the state of knowledge before and after the potential application of a sensor to a given task. For example, considering the track update allocation option, the usefulness, or the utility, of the estimate of track position can be measured by the Kalman covariance matrix. Then, the utility of track update can be computed from the present covariance matrix and the computed covariance matrix if an observation was received. For example, a typical adaptive filtering scheme will operate by increasing the Kalman filter covariance matrix upon maneuver detection. Then, the utility theory allocation method will automatically favor the update of maneuvering targets because of the anticipated utility gain associated with decreasing the covariance matrix. Similarly, the utility of search can be determined from the expected value of the targets to be detected in search.

7.2 Sensor allocation using an expert system

The use of expert system techniques for sensor allocation offers several important advantages. First, a rule-based system readily allows for expansion through the inclusion of additional rules. The use of rules also makes it easier to interpret and to explain the behavior of an expert system, as opposed to analyzing the properties of utility functions. Finally, the expert system technique conveniently allows for sensor allocation to be closely related to situation assessment, since concepts such as target hostile intent are an integral part of both the situation assessment and sensor allocation functions.

As an example of the expert system approach to sensor allocation, the method of Reference 12 starts by defining quantitative (root) concepts such as distance, closing rate, and target bearing which are determined from encounter geometry. Following this, qualitative (higher order) concepts, such as target identification (ID) and danger posed by the target, are defined. The membership of each target in the "fuzzy set" defined by these concepts is computed. For example, a simple relationship of time-to-go ($R/\dot{R}$) can be used for the root concept "close." Then using a set of "IF-THEN" rules the membership of each target in the fuzzy set defined by higher order concepts, such as dangerous, is computed. Finally, using a combination of root and higher order concepts, the membership, as well as priorities, in the set of tracks to be updated is computed and allocation of sensors is made.

8. CONCLUSIONS

It is anticipated that future increased computational capabilities will allow the use of improved methods for data association and estimation in order to effectively utilize the wide variety of data that will be available in multiple sensor systems. This means that the all-neighbors and deferred decision methods, such as MHT, should replace the current emphasis upon sequential nearest-neighbor methods. Similarly, the many advantages of Kalman filtering should make the use of fixed-gain filtering approaches essentially obsolete for future systems. Finally, the use of Bayesian and Dempster-Shafer techniques for target identification should replace the more ad hoc methods presently in common practice.

In order to effectively utilize the capability of multiple sensor systems requires an efficient allocation method. Also there will be an increasing use of artificial intelligence techniques for interpreting the track file information. Thus, it is likely that expert systems approaches will be widely applied for sensor allocation.

REFERENCES

1. A. Farina and F.A. Studer, Radar Data Processing, Wiley, New York (1985).
2. S.S. Blackman, Multiple Target Tracking with Radar Applications, Artech House, Dedham, MA (1986).
3. Y. Bar-Shalom and T.E. Fortmann, Tracking and Data Association, Academic Press, San Diego, CA (1987).
4. R.J. Fitzgerald, "Development of Practical PDA Logic for Multi-Target Tracking by Microprocessor", Proc. of 1986 American Control Conference, pp. 889-898 (1986).
5. D.B. Reid, "An Algorithm for Tracking Multiple Targets", IEEE Trans. on Automatic Control, Vol. AC-24, pp. 843-854 (1979).
6. G.V. Trunk and J.D. Wilson, "Association of DF Bearing measurements with Radar Tracks", IEEE Trans on Aerospace and Electronic Systems, vol. AES-23, pp. 438-447 (1987).
7. Y. Bar-Shalom and L. Campo, "The Effect of Common Process Noise on the Two-Sensor Fused-Track Covariance", IEEE Trans. on Aerospace and Electronic Systems, Vol. AES-22, pp. 803-805 (1986).

8. D.V.Stallard, "An Angle-Only Tracking Filter in Modified Spherical Coordinates", Proceedings of 1987 AIAA Guidance, Navigation and Control Conference, pp. 542-550 (1987)
9. P.J. Nahin and J.L. Pokoski, "NCTR Plus Sensor Fusion Equals IFFN or Can Two Plus Two Equal Five?", IEEE Trans. on Aerospace and Electronic Systems, Vol AES-16, pp. 320-337 (1980)
10. P.L. Bogler, "Shafer-Dempster Reasoning with Applications to Multi-sensor Target Identification Systems", IEEE Trans. on Systems, Man and Cybernetics, Vol. SMC-17, pp. 968-977 (1987).
11. G.W. Deley, "A Netting Approach to Automatic Radar Track Initiation, Association and Tracking in Air Surveillance Systems", AGARD Conference Proceedings No. 252, Strategies for Automatic Track Initiation, Monterey, CA, pp. 7-1 to 7-10 (1978).
12. R. Popoli and S. Blackman, "Expert System Allocation for the Electronically Scanned Antenna Radar", Proceedings of 1987 American Control Conference, pp. 1821-1826 (1987).

Information fusion methodology

Gerald M. Flachs, Jay B. Jordan, and Jeffrey J. Carlson

New Mexico State University
Department of Electrical and Computer Engineering
Box 3-0, Las Cruces, New Mexico 88003

ABSTRACT

An approach is presented for designing multisensor electronic vision systems using information fusion concepts. A random process model of the multisensor scene environment provides a mathematical foundation for fusing information. A complexity metric is introduced to measure the level of difficulty associated with various vision tasks. This complexity metric provides a mathematical basis for fusing information and selecting features to minimize the complexity metric.

A major result presented in the paper is a method for utilizing a priori knowledge to fuse an n-dimensional feature vector $X = \{X_1, X_2, ..., X_n\}$ into a single feature Y while retaining the same complexity. A fusing theorem is presented that defines the class of fusing functions that retains the minimum complexity.

1. INTRODUCTION

Electronic vision systems are called upon to assimilate multiple sources of information into intelligent control decisions. These systems must be capable of analyzing complex multisensor scenes, locating potential objects of interests, segmenting objects from cluttered backgrounds, classifying objects into different object classes, and performing intelligent control decisions. These tasks are often referred to as the cuer, segmenter, recognizer, and controller tasks, respectively. The analysis and design of an electronic vision system is concerned with measuring and improving the system's performance with respect to each of these tasks as it operates in various scene environments.

Current attempts to improve the performance of such systems are based upon incorporating multiple sources of information into the decision process. These newer systems often demonstrate significantly improved performance by fusing the different information sources to obtain better scene descriptions and control decisions. The recent success of these systems motivates the development of a theoretical foundation for information fusion. To obtain a theoretical basis for information fusion, a mathematical model of the scene environment is needed to characterize the environment in which the vision system operates. The environment includes the sensor characteristics, atmospheric effects, and the scene dynamics. With a mathematical model of the scene environment, a complexity metric is introduced to measure the level of difficulty associated with various vision tasks in a given scene environment. The complexity metric can be used as a basis for feature selection and to develop functions for fusing multiple sources of information into a simpler description without increasing the complexity of the tasks. The information fusion problem is then presented as a process of assimilating multiple information sources into a single composite information source followed by an optimal decision process to select the control strategy. The information fusion and decision processes utilize knowledge of physical laws to establish the fusing function and a priori or gained knowledge to form an optimal control decision.

2. MULTISENSOR SCENE ENVIRONMENT and MULTISENSOR IMAGES

Electronic vision systems perceive their environment with sensors that take measurements in a multidimensional electromagnetic energy field. This field has a spectral, a temporal, and three spatial dimensions. The electromagnetic energy observed originates from numerous emitters in the three-space some of which may be present on the imaging system (RADAR, LADAR, etc.). The energy from the various emitters is reflected, absorbed, and re-radiated from *objects* and *background* in the observation space. The *imaging system* consists of one or more sensors that convert the average energy (i.e. **E** × **H**, Poynting Vector) in relatively small regions of space, spectra, and time into electrical signals for processing by both analog and digital electronic devices. A goal, typical of many electronic vision systems, is to deduce from these signals the presence and position of entities termed *targets*. The performance of the system is measured by how well it minimizes the errors of classifying *nontargets* or *background* as *targets* (*false alarms*) and classifying *targets* as *background* (*misses*)[1].

The term *scene* is used to describe specific signals (both analog and digital) derived from the energy field by an electromagnetic energy sensing device and its associated electronics. The values of these signals at any point in time are influenced by a myriad of factors determining the state of the energy field and physical processes in the sensor and its associated electronics.

If the overall effect of all unpredictable factors on the observed signals is considered *random*, the signals composing the *scene* are modeled as a stochastic or random process[1][2][3] in time, space and spectra. This process is the model proposed for the *scene environment*. For multiple sensor systems, there is a random process associated with each sensor. At any point in time, region of space and spectra, a sensor observes an instance of a single realization of the process. That is, it observes the values of a set of *random variables*. The term *image* is used to describe this set of observations the sensor obtains from the random process. The set of sensors and their associated electronics is termed the *imaging system*.

Figure 1. presents a conceptual overview of the statistical scene environment model. The spatial coordinates of the energy field are denoted x,y,and z. The temporal and spectral coordinates are denoted t and s respectively. If we let Z represent the tuple (x,y,z,t,s), then the portion of the energy field observed by the sensor is a random process $E(Z) \times H(Z)$ with random factors arising from such things as atmospheric turbulence, particulate activity, fluctuations in energy sources, etc. Each realization of the random process modeling the energy field is a five dimensional continuous space. Each point in the space is a single observation of a vector random variable. As with all physical processes modeled as random processes, increasing the degree of knowledge about the underlying deterministic physical properties decreases the amount of randomness.

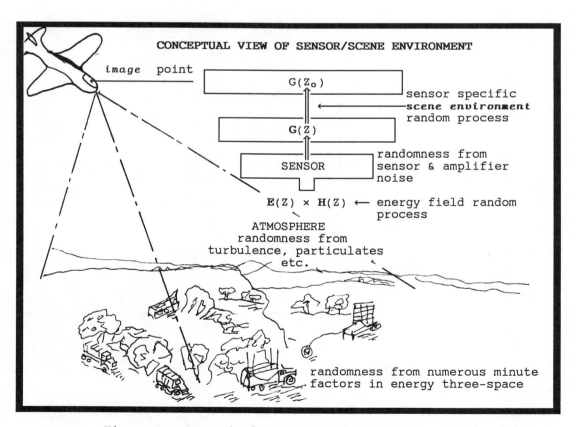

Figure 1. Conceptual Overview of Scene Environment

The sensor transforms a realization of the random process $E(Z) \times H(Z)$ to another realization $G(Z)$ in which further random factors such as sensor and amplifier noise are introduced. The sensor transformation also includes averaging (often nonlinear) over space, time, and spectra. (Note that most sensors are designed to sense the direction of $E(Z) \times H(Z)$ as well as its magnitude.) In this case, the random process $G(Z)$ is not only a function of position, but of sensor attitude as well. In this case, Z represents the tuple $(x,y,z,\theta,\phi,\psi,t,s)$. Where θ,ϕ, and ψ are the sensor attitude variables roll, pitch, and yaw.

An imaging sensor, such as a focal plane infrared (IR) detector array, may be considered a multisensor which generates a realization of a different random process $G_{ij}(Z)$ for each of the sensor IR detector elements, where i and j denote the row and column indices of each detector element respectively. For any given set of coordinates Z_0 the $G_{ij}(Z_0)$ form an observation from a two dimensional array of random variables. A set of single observations (a sample) from the array of random variables forms a *digital image* observed by the system. The basic concept of a random process for each sensor element is naturally extended to describe the scene environment of systems with multiple sensor types and geometries.

In general, the joint distribution of the random variables giving rise to the observed image is incredibly complex. However, the random process describing the scene environment often contains regions in space, time and spectra with relatively simple statistical properties[4]. The existence of such regions provides the fundamental justification for the metrics and techniques presented.

2.1 TARGET/BACKGROUND DICHOTOMY

The target/background dichotomy representation of a scene environment that arises in the context of many vision systems is an implicit statement of a priori information concerning the random processes modeling the scene environment. Specifically, the *a priori knowledge* that there are two classes of objects, *targets* and *nontargets* allows us to partition the set of random variables comprising the random process $G(Z)$ into two subsets described by two random processes $G_T(Z_T)$ and $G_B(Z_B)$, where the domain of variables $Z_T = (x_T,y_T,z_T,t_T,s_T)$ define the region of space, time, and spectra in the energy field containing the *target* signature. The process, $G_B(Z_B)$, describing the remainder of the energy field is considered to contain only the signatures of background and/or false targets.

In terms of the random process model of the scene environment, the tasks of detecting the target in the background and separating the target from the background are naturally posed as statistical hypothesis tests to determine which of the sets of random variables gave rise to the observed measurement(s). Formulation of the hypothesis test requires that the joint distributions of the random variables composing the targets and background somehow be estimated[5].

The target *recognition* problem further partitions the set of random variables comprising G_T into several subsets of random variables comprising processes G_{T1}, G_{T2}, ... G_{Tn} describing each class of targets. The problem of identifying or classifying the target is an m-ary hypothesis test to determine which of the target classes gave rise to the observed value(s). Again, formulation of this test requires the estimation of the joint distributions of the random variables composing each target class.

This statistical decision making technique is conceptually simple, however, its application in practice is extremely difficult. The reason is that for each scene, there is a *continuum* of random variables in the random process describing the target. Likewise, there is a *continuum* of random variables in the random process describing the background. It is not practical in real scenes to acquire enough information to fully characterize the incredibly complex joint distributions of the random variables composing the random processes. For this reason, it is necessary to use some function (or set of functions) of the random variables that transforms the *continuum* to a finite set of random variables, $\{X_1,X_2,...,X_n\}$, called *features*. As the set is ordered, it is often denoted $X = [X_1,X_2,...,X_n]$ where X is termed the feature vector. This reduction in dimensionality of the problem[6], is essential for the implementation of an actual system.

To illustrate the statistical decision making process, consider the target locating and segmenting tasks with random variables $X_1,X_2,...,X_n$, to be features common to both target and background. The set $x = \{x_1, x_2, ..., x_n\} = [x_1, x_2, ..., x_n]$ is the observed feature vector, $f(x_1,x_2,...,x_n|T) = f(x|T)$ is the conditional joint probability density or mass function (pdf) of the random variables under the target hypothesis, and $f(x_1,x_2,...,x_n|B) = f(x|B)$ is the conditional joint pdf of the random variables under the background hypothesis.

The typical hypothesis test, based on observed values $x_1, x_2, \ldots x_n$ is stated:

$$H_0(T): x_1, x_2, \ldots, x_n \text{ are observations from } f(x_1, x_2, \ldots, x_n | T)$$
$$H_1(B): x_1, x_2, \ldots, x_n \text{ are observations from } f(x_1, x_2, \ldots, x_n | B)$$

Decisions to classify the observations as being from target (H_0) or background (H_1) are typically made in such a manner that either a) the probability of making an error is minimized or b) the cost of making an error is minimized[1]. A similar set of tests is constructed for the classification of targets into various target classes.

2.2 FEATURE SELECTION AND EXTRACTION

There are several important things to note about the transformation from **scene space** to **feature space**. First, the features used in a system are **task dependent**. The recognition task does not necessarily use the same set of features as the detection or segmentation tasks. In general, features can be derived from any spatial, temporal, and spectral transformation of the observation space. The simplest of systems use just a single feature (gray-level relative frequency histograms from IR imagery, for example). Preliminary experiments with the complexity metric, presented in the next section, indicate that the performance of a system may depend more on the features it uses than on any other factor.

3. COMPLEXITY METRIC

A complexity metric[9] is introduced to measure the level of difficulty associated with vision tasks. Four major tasks are to locate potential objects of interest (**targets**), separate them from the background scene, classify objects into various **target classes**, and make intelligent control decisions. The complexity metric provides a mathematical basis for measuring how difficult each task is to perform in a given scene environment.

Complexity is defined in terms of a target/background dichotomy, a vector of random variables $X = (X_1, X_2, \ldots, X_n)$ representing features used in the vision task that are common to both target and background, and the conditional probability density functions $f(x|T)$ and $f(x|B)$. If the random variables are continuous then complexity is defined by

$$\mathbb{C}(x) = \int_{-\infty}^{\infty} f(x|T) \wedge f(x|B) dx$$

where the operator $\wedge$ is the minimum operator. If the random variables are discrete, then complexity is defined in terms of the conditional joint probability mass functions.

$$\mathbb{C}(x) = \sum_{\forall x} f(x|T) \wedge f(x|B).$$

To illustrate the complexity metric graphically, consider a single feature, X, common to both target and background. In each graph, the hashed area corresponds to the degree of complexity under that feature. The first graph represents a degree of complexity that is much less than that shown in the second graph. The least level of complexity occurs when the density functions under each hypothesis are completely disjoint. In this case complexity is zero. The greatest level of complexity occurs when the density functions under each hypothesis totally overlap. In this case complexity is one.

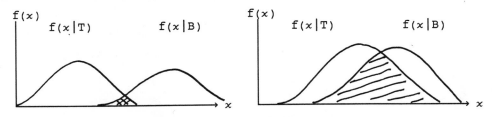

Figure 2. Two Levels of Complexity

The following properties of the complexity metric are a direct result of the fundamental definition.

1) $0 \leq \mathbb{C} \leq 1$.
2) $\mathbb{C} = 0$ implies no common feature measurements between target and background.
3) $\mathbb{C} = 1$ implies all common feature measurements in the same proportions.

An important characteristic of the complexity metric is its direct relationship to the minimum probability of error (MPE) associated with a two class decision rule. When the apriori probabilities of each decision class are equal, complexity equals twice the MPE.

Theorem

If the a priori probabilities of two decision classes T and B are equally likely, $P(T) = P(B) = 0.5$, then $\mathbb{C} = 2$ MPE.

Proof

The probability of error associated with a two class decision rule is given as:

$$P_e = P(T)\int_{\mathbb{R}_2} f(x|T)dx + P(B)\int_{\mathbb{R}_1} f(x|B)dx \tag{1}$$

where:

$\mathbb{R}_1 = \{x: x \text{ is classified as belonging to } T\}$ (2)
$\mathbb{R}_2 = \{x: x \text{ is classified as belonging to } B\}$ (3)

The decision rule that minimizes the probability of error is the liklihood ratio test[1].

$$\frac{f(x|T)}{f(x|B)} \underset{B}{\overset{T}{\gtrless}} \frac{P(B)}{P(T)} \ . \tag{4}$$

Equivalently,

$$P(T)f(x|T) \underset{B}{\overset{T}{\gtrless}} P(B)f(x|B). \tag{5}$$

For a minimum probability of error decision rule, $\mathbb{R}_1$ and $\mathbb{R}_2$ are restricted as follows:

$\mathbb{R}_1 = \{x: P(T)f(x|T) > P(B)f(x|B)\}$, and (6)
$\mathbb{R}_2 = \{x: P(T)f(x|T) \leq P(B)f(x|B)\}$. (7)

The restriction on $\mathbb{R}_1$ and $\mathbb{R}_2$ for minimizing the probability of error results in the following compact expression for the MPE.

$$MPE = \int_{-\infty}^{\infty} P(T)f(x|T) \wedge P(B)f(x|B)dx. \tag{8}$$

Letting $P(T) = P(B) = 0.5$ leads to the desired result.

$$MPE = 0.5 \int_{-\infty}^{\infty} f(x|T) \wedge f(x|B)dx = 0.5\mathbb{C} \tag{9}$$

The complexity metric gains technical merit from its relationship to the Kolmogorov distance (KD). Under certain conditions, $\mathbb{C} = 1-KD$, and the distribution of the complexity metric can then be expressed in terms of the known distribution of the KD.

Theorem

If $f(x|T)$ and $f(x|B)$ have a single intersection point and $x \in R^1$ then $\mathbb{C} = 1-KD$.

Proof

Define a distance function

$$D(x) \triangleq \left[\int_{-\infty}^{x} f(z|T) - f(z|B)dz \right]^2, \tag{10}$$

and KD, the maximum of $\sqrt{D(x)}$.

The condition for D(x) to be maximized is obtained by differentiating D(x) with respect to x and equating the result to zero. The maximum occurs at f(x|T) = f(x|B), the intersection point. Now consider the following diagram.

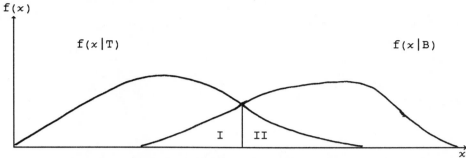

Figure 3. Complexity and the Kolmogorov Distance

Complexity and the KD can be expressed in terms of the areas denoted by I and II.

    ℂ = I + II.
    KD = (1 - II) - I.
    KD = 1 - (I + II).
    KD = 1 - ℂ.

    Hence,   ℂ = 1 - KD.

4. AN INFORMATION FUSION PROCESS

The random process model of the scene environment and the complexity metric provide a mathematical basis for fusing multiple sources of information into optimal control decisions. The concept of an information source is used to represent any source of information that is available to the vision system. It may be a direct output of a sensor or it may be derived from a sensor signal by some feature computational process such as edge filter, motion detector, or texture measurement. Consequently, a single sensor can be viewed as a multiple information source when several features are extracted and used in the fusion process.

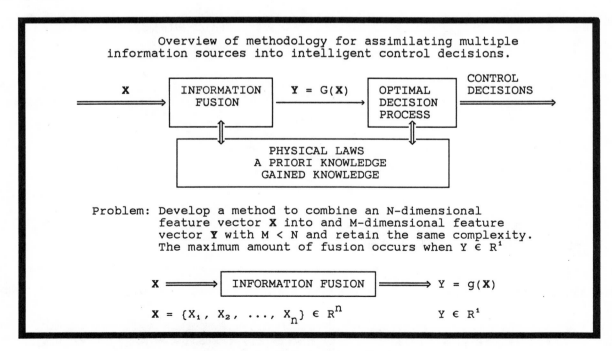

Figure 4. Information fusion.

Figure 4. illustrates the technique of fusing multiple information sources into a single composite information source followed by an optimal decision process. Letting **X** represent an N-dimensional vector of information sources, and Y represent the single composite information source, the information fusion problem is reduced to finding a function, Y = g(**X**), to combine the information and retain the same complexity, (keeping $\mathbb{C}(\mathbf{X}) = \mathbb{C}(Y)$). In preserving complexity, the level of difficulty in performing a vision task does not increase.

A fusion theorem is presented to define the class of functions that retain complexity. Let $f(\mathbf{x}|T)$ and $f(\mathbf{x}|B)$ be the conditional joint probability density functions for a two class decision problem of deciding target (T) or background (B). Further define:

$$\mathbf{X} = \{X_1, X_2, \ldots, X_n\} \in R^n.$$

$$\mathbb{R}_T = \left\{ \{x_1, x_2, \ldots, x_n\} \mid f(\mathbf{x}|T) > f(\mathbf{x}|B) \right\}. \tag{11}$$

$$\mathbb{R}_B = \left\{ \{x_1, x_2, \ldots, x_n\} \mid f(\mathbf{x}|B) > f(\mathbf{x}|T) \right\}. \tag{12}$$

$$\mathbb{R}_E = \left\{ \{x_1, x_2, \ldots, x_n\} \mid f(\mathbf{x}|T) = f(\mathbf{x}|B) \right\}. \tag{13}$$

$$Y = g(\mathbf{X}) \in R^1$$

$$\mathbb{R}(y) = \left\{ (x_1, x_2, \ldots, x_n) \mid y = g(x) \right\} \tag{14}$$

<u>Fusion Theorem</u>[7]. $\mathbb{C}(\mathbf{X}) = \mathbb{C}(Y)$ iff for each y either $\mathbb{R}(y) \cap \mathbb{R}_B = \phi$ or $\mathbb{R}(y) \cap \mathbb{R}_T = \phi$.

Proof: Consider the discrete complexity definition

$$\mathbb{C}(\mathbf{X}) = \sum_{\forall x} f(\mathbf{x}|T) \wedge f(\mathbf{x}|B) \tag{15}$$

and define:

$$\mathbb{R}_{T(y)} = \left\{ y \mid \mathbb{R}(y) \cap \mathbb{R}_B = \phi, \; \mathbb{R}(y) \cap \mathbb{R}_T \neq \phi \right\}, \tag{16}$$

$$\mathbb{R}_{B(y)} = \left\{ y \mid \mathbb{R}(y) \cap \mathbb{R}_T = \phi, \; \mathbb{R}(y) \cap \mathbb{R}_B \neq \phi \right\}, \text{ and} \tag{17}$$

$$\mathbb{R}_{E(y)} = \left\{ y \mid \mathbb{R}(y) \cap \mathbb{R}_T = \phi, \; \mathbb{R}(y) \cap \mathbb{R}_B = \phi \right\}. \tag{18}$$

Now write the complexity in terms of the partitioning $\mathbb{R}_{T(y)}$, $\mathbb{R}_{B(y)}$, and $\mathbb{R}_{E(y)}$.

$$\mathbb{C}(\mathbf{X}) = \sum_{\forall y} \sum_{\mathbb{R}(y)} f(\mathbf{x}|T) \wedge f(\mathbf{x}|B). \tag{19}$$

$$\mathbb{C}(\mathbf{X}) = \sum_{y \in \mathbb{R}_{T(y)}} \sum_{\mathbb{R}(y)} f(\mathbf{x}|T) \wedge f(\mathbf{x}|B) + \sum_{y \in \mathbb{R}_{B(y)}} \sum_{\mathbb{R}(y)} f(\mathbf{x}|T) \wedge f(\mathbf{x}|B) + \sum_{y \in \mathbb{R}_{E(y)}} \sum_{\mathbb{R}(y)} f(\mathbf{x}|T) \wedge f(\mathbf{x}|B). \tag{20}$$

Observe that

$$f(\mathbf{x}|T) > f(\mathbf{x}|B) \quad \text{if } y \in \mathbb{R}_{T(y)},$$
$$f(\mathbf{x}|B) > f(\mathbf{x}|T) \quad \text{if } y \in \mathbb{R}_{B(y)}, \text{ and}$$
$$f(\mathbf{x}|B) = f(\mathbf{x}|T) \quad \text{if } y \in \mathbb{R}_{E(y)}.$$

Hence:

$$\mathbb{C}(\mathbf{X}) = \sum_{y \in \mathbb{R}_{T(y)}} \sum_{\mathbb{R}(y)} f(\mathbf{x}|B) + \sum_{y \in \mathbb{R}_{B(y)}} \sum_{\mathbb{R}(y)} f(\mathbf{x}|T) + \sum_{y \in \mathbb{R}_{E(y)}} \sum_{\mathbb{R}(y)} f(x|T). \tag{21}$$

Since $\mathbb{R}(y) \cap \mathbb{R}_T = \phi$ or $\mathbb{R}(y) \cap \mathbb{R}_B = \phi$ for each y is a necessary and sufficient condition for interchanging the summation $\sum_{\mathbb{R}(y)}$ with the minimum operator $\wedge$, complexity can be written

$$\mathbb{C}(\mathbf{X}) = \sum_{y \in \mathbb{R}_{T(y)}} \left[\sum_{\mathbb{R}(y)} f(\mathbf{x}|T) \right] \wedge \left[\sum_{\mathbb{R}(y)} f(\mathbf{x}|B) \right] + \sum_{y \in \mathbb{R}_{B(y)}} \left[\sum_{\mathbb{R}(y)} f(\mathbf{x}|T) \right] \wedge \left[\sum_{\mathbb{R}(y)} f(\mathbf{x}|B) \right] + \sum_{y \in \mathbb{R}_{E(y)}} \left[\sum_{\mathbb{R}(y)} f(\mathbf{x}|T) \right] \wedge \left[\sum_{\mathbb{R}(y)} f(\mathbf{x}|B) \right].$$

$$\tag{22}$$

Hence

$$\mathbb{C}(\mathbf{X}) = \sum_{\forall y} \left[\sum_{\mathbb{R}(y)} f(\mathbf{x}|T) \right] \Lambda \left[\sum_{\mathbb{R}(y)} f(\mathbf{x}|B) \right] = \sum_{\forall y} f(y|T) \Lambda f(y|B) = \mathbb{C}(\mathbf{Y}). \qquad (23)$$

Based upon the fusion theorem there are many functions that can be used to combine multiple information sources into a composite information source. Some interesting complexity perserving fusing functions $Y = g(\mathbf{X})$ include:

$Y = f(\mathbf{x}|T) - f(\mathbf{x}|B)$

$Y = f(\mathbf{x}|T) / f(\mathbf{x}|B)$

$$Y = \begin{cases} 1 & f(\mathbf{x}|T) > f(\mathbf{x}|B) \\ 0 & f(\mathbf{x}|T) = f(\mathbf{x}|B) \\ -1 & f(\mathbf{x}|B) > f(\mathbf{x}|T) \end{cases}$$

All of these functions require the conditional joint probability density functions. These pdf functions can be derived from physical laws[8], or they can be experimentally estimated using observed or gained knowledge.

The likelihood ratio function, $Y = f(\mathbf{x}|T)/f(\mathbf{x}|B)$, has been shown to provide an optimal decision fusing function[1] for minimal probability of error decisions.

5. CONCLUSIONS

This paper presents initial efforts toward establishing a mathematical basis for fusing multiple sources of information into control decisions. The scene environment model, the complexity metric, and the fusion theorem provide a mathematical basis for information fusion in electronic vision systems.

6. ACKNOWLEDEMENTS

The research reported in this paper was sponsored by U.S. Army Research Office, Research Triangle Park, NC, Contract No. DAAL03-87-K-0106 and the U.S. Army Vulnerability Assessment Laboratory, White Sands Missile Range, NM, Contract No. DAAD07-83-C-0031.

7. REFERENCES

1. Van Trees, H., *Detection, Estimation, and Modulation Theory*, Wiley, 1971.

2. Papoulis, A., *Probability, Random Variables, and Stochastic Processes*, McGraw-Hill, 1965.

3. Srinath, M. and Rajasekaran, P., *An Introduction to Statistical Signal Processing with Applications*, Wiley, 1979.

4. Jordan, J. and Flachs, G. "Statistical Segmentation of Digital Images," SPIE, Vol. 754, No. 34, 1987.

5. Bendat, J. and Piersol, A., *Measurement and Analysis of Random Data*, Wiley, 1966.

6. Duda, R. and Hart, P., *Pattern Classification and Scene Analysis*, Wiley 1974.

7. Rogers, G., Nguyen, H., Walker, A., Zhang, L., Durack, D., and Flachs, G., "Communications of the Statistics Seminar at New Mexico State University," 1987.

8. Shapiro, J. H., Reinhold, R. W., and Park, D., "Performance Analysis for Peak-Detecting Laser Radars," Proceedings SP1I, Vol. 663, pp. 38-58, 1986.

9. Carlson, J., "Task Specific Complexity Metrics for Electronic Vision," Ph.D. dissertation, New Mexico State University, 1988.

An Alternate View of Munition Sensor Fusion

J. R. Mayersak, PhD

Assistant to the President for Science and Technology
Textron Defense Systems
201 Lowell Street, Wilmington, Massachusetts 01887

ABSTRACT

An alternate multimode sensor fusion scheme is treated. The concept is designed to acquire and engage high value relocatable targets in a lock-on-after-launch sequence. The approach uses statistical decision concepts to determine the authority to be assigned to each mode in the acquisition sequence voting and decision process. Statistical target classification and recognition in the engagement sequence is accomplished through variable length feature vectors set by adaptive logics. The approach uses multiple decision for acquisition and classification, in the number of spaces selected, is adaptively weighted and adjusted. The scheme uses type of climate -- arctic, temperate, desert, and equatorial -- diurnal effects -- time of day -- type of background, type of countermeasures present -- signature suppresssion or obscuration, false target decoy or electronic warfare -- and other factors to make these selections. The approach is discussed in simple terms. Voids and deficiencies in the statistical data base used to train such algorithms is discussed. The approach is being developed to engage deep battle targets such as surface-to-surface missile systems, air defense units and self-propelled artillery.

1. INTRODUCTION

The potential of precision guided and sensor fuzed lock-on-after-launch munitions to offer an order of magnitude increase in capability compared to existing inventory munitions has been documented. Full scale engineering development programs are in process in both the Army and the Air Force focused on such munitions. Many of these concepts employ more than one sensor, in combination, to discriminate the target in the presence of competitive returns from the background clutter. Dual mode systems, representing the simplest of these implementations, require the designer to consider the fusion of data from the two sensors -- infrared and millimeter wave, infrared and active laser radar, and millimeter wave and active laser radar -- in performing the required functions of target acquisition, aimpoint selection, guidance and control, and countermeasure suppression.

2. BACKGROUND

The task confronting the munition designer in considering the fusion of the outputs of various sensors is complex. The munition designer has to operate under specific constraints including:

a. Cost. A multimode sensor system has to be implemented within specified cost thresholds. Generally, these thresholds are to be taken as less than $5,000 per munition, if the munition is sensor fuzed, and $25,000 if the munition is precision guided.

b. Complexity. The complexity of the fusion logic is generally limited to those data, for use in setting weighted thresholds and assigning authority in the decision process, which is available through the munition systems. Weapon platform designers are extremely reluctant to hand data over from other sensor systems located on the delivery platform to the munition systems.

The shortage associated with the design data base to support the definition of fusion schemes for sensor fuzed munitions presents a significant problem to the munition designer. These shortages include:

a. Signature Data. The training of any sensor fusion decision logic is limited by the unavailability of:

(1) High Value Fixed Target Data. The availability of signature data to support the design of a multimode munition decision logic is limited. The ability of the designer to access signature data for aircraft shelters, POL facilities, bunkers, bridges, power plants, and other high value fixed targets where direct measurements is limited.

(2) Mobile Force. The availability of signature data for current inventory and future enemy medium tanks, self-propelled artillery, surface-to-surface missile launchers, transporter loaders, air defense units and other mobile force targets is limited. (Figure 1)

b.   Countermeasure Data. The availability of data describing probable enemy countermeasures is limited. (Figure 2) This would include:

VOIDS AND DEFICIENCIES

- TARGET SIGNATURES
  - CURRENT
    - MEDIUM TANKS
      - T-64
      - T-72
    - SELF PROPELLED ARTILLERY
      - M-1973
      - M-1974
    - SURFACE-TO-SURFACE MISSILES
    - AIR DEFENSE UNITS
  - FUTURE
    - MEDIUM TANKS
      - T-80
      - FS-T
    - SELF PROPELLED ARTILLERY
      - FS-SPH
    - SURFACE-TO-SURFACE MISSILES
      - SS-23
      - FS-SSM
    - AIR DEFENSE UNITS
      - SS-12
      - SS-13
      - FS-ADU

46-4856

Figure 1

CRITICAL ISSUES

- COUNTERMEASURES
  - SIGNATURE REDUCTION
    - OBSCURATION
    - SUPPRESSION
    - CONCEALENT
    - CAMOUFLAGE
  - FALSE TARGET DECOY
    - SIMPLE
    - COMPLEX
      - HIGH FIDELITY
  - LETHAL STRIKE
    - CONVENTIONAL
    - DIRECTED ENERGY
      - HIGH ENERGY LASER
      - MICROWAVE
  - ELECTRONIC WARFARE
    - JAMMERS
    - GATE STEALING REPEATERS
    - TRACK DISRUPTION
      - DIRECTED ENERGY
  - TACTICS
    - DECEPTION
      - CO-LOCATED DECOY
      - MULTIPLE SHELTER
    - JOINT CONCEPTS
      - COMBINATION

46-6085

Figure 2

(1) Signature Reduction. The ability to reduce the target signature, through obscuration or suppression to reduce the probability of the vehicle being acquired by surveillance and acquisition systems or by munition systems or to shift the aimpoint on the vehicle for precision guided and sensor fuzed munitions to a less vulnerable area or off the vehicle, is a concern.

(2) False Target Decoy. The ability to use false target decoys to reduce the vulnerability of a target being acquired due, to increasing the number of target sets offered and the ability to shift aimpoints to less vulnerable areas on the target or off the target, is a concern.

(3) Electronic Warfare. The ability to effect one or more modes in a multimode sensor suite, through electronic warfare techniques including signal processing disruption, high powered microwave upset, and directed energy weapons, is a concern.

(4) Camouflage, Concealment, and Deception. The ability to use camouflage, concealment, and deception techniques to reduce the vulnerability of high value fixed targets or mobile force targets, through techniques as simple as employing properly constructed camouflage nets, to reduce the vulnerability of these targets to the engagement and attack by precision guided and sensor fuzed munitions, is a concern.

(5) Tactics. The ability of the enemy to use combined sets of signature reduction, false target decoy, electronic warfare and camouflage and concealment and deception in a well planned and executed countermeasure program is a concern.

The principle reason for the inclusion of more than one sensor in a multimode sensor fuzed or precision guided munition may be the additional discrimination capability offered by the additional sensors in discriminating countermeasure techniques in the engagement sequence.

c. Climactic Variation. Multimode systems offer the ability of the munition to discriminate high value or mobile force target in a variety of climactic backgrounds -- arctic, temperate, desert, and equatorial. The ability of the multimode sensor suite to discriminate the presence of the target differs in each background type. The sensor fusion scheme, for example, would assign greater authority to the infrared channel in a dual mode infrared millimeter wave system when operating in an arctic background, due to the difficulties millimeter wave systems have in discriminating targets in the presence of refrozen or metamorphic snow. It would assign a greater role in the decision process of the millimeter wave system, over the infrared system, when operating in a temperate environment due to the increase false targets associated with infrared discrimination logics in that type of background.

d. Diurdinal Variation. The passive, and active, signatures associated with high value fixed targets and mobile force targets varies as a function of the time of day. The targets may actually experience a reversal of contrast, moving from positive contrast to negative contrast, over any 24-hour cycle. The weighting of the various sensor outputs and the voting of the acquisition decision, as well as the selection of aimpoint has to account for these variations.

e. Weather. The target signature varies with the type of weather present. The signature monitored by several modes may be effected. The infrared signature depends on solar load, cloud cover, the presence of rain, and other factors. Millimeter wave signatures vary considerably with the rate of rainfall and the moisture content in the background clutter.

All of the factors described above have to be considered by the designer in the selection of the appropriate sensor fusion process. (Figure 3) Most often, the ability to define the sensor fusion scheme far exceeds the knowledge available in the data base for the munition designer.

Options for definition of algorithms for sensor fusion (Figure 4) include:

a. Heuristic. Often the best technique for defining the voting logics and weighting the outputs of various sensors in a multisensor precision guided or sensor fuzed munition depends on past experience. The impirical or experience based decision logics based on large training sets may, in fact, though not as sound mathematically as more sophisticated approaches, provide an excellent approach.

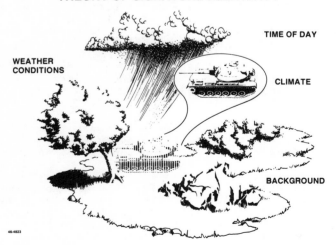

THEORY OF SIGNATURE RELATIVITY

TIME OF DAY

WEATHER CONDITIONS

CLIMATE

BACKGROUND

Figure 3

ALGORITHM TYPES

- HEURISTIC
  - EMPIRICAL/EXPERIENCE BASED
  - LARGE TRAINING SET

- NON PARAMETRIC
  - MOST GENERAL
  - POWERFUL TECHNIQUE
    - COMPUTATION INTENSE
  - FORMAL THEORY FRAGMENTED

- PARAMETRIC
  - MOST EFFICIENT
  - POWERFUL
  - FORMAL THEORY ESTABLISHED
    - STATISTICAL PROBLEM
  - FEATURE SELECTION ARBITRARY

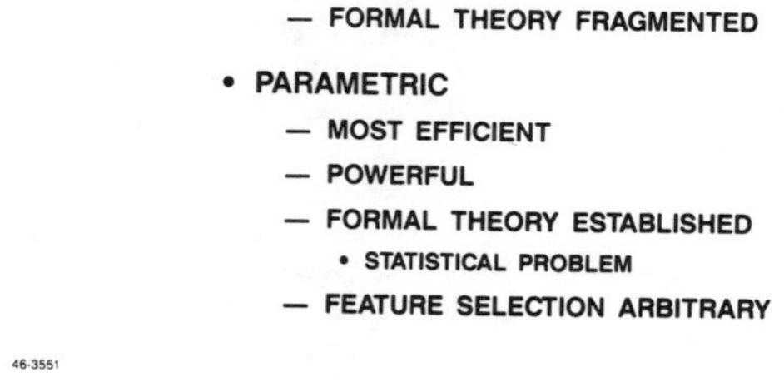

Figure 4

b. Nonparametric. Nonparametric schemes have received, in a large part, the greatest amount of attention. They represent the most potentially powerful approach available. Neighborhood transforms such as the Kth nearest neighbor, two dimensional rolling ball approximations and other neighborhood transform logics, including correlation systems, constitute the principle tools in this design approach for sensor fusion. Nonparametric schemes, combined with matched filters concepts, have demonstrated considerable authority in assigning and weighting inputs into a fusion scheme. The approach suffers from two fundamental limitations including:

(1) Fragmented Formal Theory. While the fundamental concept for neighborhood transform operations is fairly well understood and recognized, the mathematical calculus associated with these schemes has not been formally consolidated and defined. The same problem presented to several designers will result in as many solutions as there are designers. The ability to compare the authority of sensor fusion schemes and to determine which one provides the greatest return on investment, weighed against specific criterion such as computer throughput requirements, is difficult.

(2) Computationally Intense. These schemes usually are associated with computational requirements of over a million complex operations per second and can go as high as 25 million complex operations per second.

(3) Data Base. The data base against which to train these algorithms, in general, is not available. The training of even the most complex and mathematically sophisticated fusion scheme often fails on the quality of the data against which to set the decision criterion.

c. Parametric. The parametric scheme, using statistically based algorithms, represent a sensor fusion approach which makes target acquisition and aimpoint selections, in the voting logic of the sensor fusion scheme, based on the information available. The statistical decision process employs classical statistical theories such as:

(1) Fischer Linear Statistics. The technique allows separation of the target and background data ensembles when covariant matrices of the target and clutter are essentially identical.

(2) Karhunen-Loeve. This technique offers the ability to discriminate a target from clutter when the clutter ensemble cannot be sampled and operates by forming a minimal acceptance volume for the target distribution ensemble in the n-dimensional space defined by the features in the discrimination set.

d. Quadratic Bayesian Statistics. This technique allows the separation of the target and background data ensembles where the target and clutter covariant matrices are known and the distribution associated with the target and the clutter are Gaussian.

The principle reason for the use of statistically parametric based algorithms is found in consideration of the data base. The data base available to define the statistical ensembles including the mean and covariance matrices in the n-dimensional space in which the decision processes for each of the multi modes and the voting logic in the sensor fusion schemes are made, is limited. While all of the sensor fusion schemes described have advantages and disadvantages in any application, it can be argued in those applications where the data base is limited the parametric scheme offers the most reasonable approach for a multimode sensor fusion scheme. In such a mode, the features in any sensor channel would be selected to provide the maximum separation in the statistical ensembles associated with the target and background clutter to allow the statistical process of the nearest mean determination to be employed with considerable authority. The weighting of the inputs associated with any sensor channel would be accomplished by weighting the features or changing the features in the acquisition process. The fusion process in the determination of the target acquisition could, then, be as simple as a logic and/or decision. It may be as complicated as a threshold scoring with the decision of each channel considered in a conditional probability calculation.

3. BASIC APPROACH

To understand the sensor fusion process, consider a simple case involving the probability distribution functions for two different ensembles -- a set of black balls and a set of white balls -- contained in a box. Assume the diameter of each type of ball to be governed by a Gaussian probability distribution. (Figure 5) The diameter of the balls is a feature. Assume the black balls are associated with the frequency distribution having the larger mean value. The ability to identify either a black ball or a white ball drawn from the box in a darkened room will depend on the ability of that feature -- the diameter of the ball -- to discriminate between the black balls and the white balls. In a formal sense, the feature would be selected to ensure maximum separation of the means of the two statistical ensembles. If the diameter of the ball drawn from the box is extremely small, it can be said to be most likely a white ball. If the diameter of the ball is extremely large, it can be said to be most likely a black ball. For that diameter which is separated in an equal number of black ball standard deviations from the black ball mean and the same number of white ball standard deviations from the white ball mean, it is impossible to say that the ball is more likely to belong to the black or white ball ensemble. It is equally likely that it could be a member of either data ensemble. To the right of this equalilikelihood point, it is more likely that the ball is a member of the black ball ensemble. To the left of this point is more likely the ball is a member of the white ball data ensemble. A ball drawn to the right of the equalilikelihood point has as its nearest mean the black ball ensemble mean. To the left of the equalilikelihood point the nearest mean is the white ball ensemble. As a consequence, the decision process is associated with the determining the nearest mean.

As the process of drawing balls from the box in a darkened room continues, balls drawn near the equalilikelihood point will be incorrectly identified from time to time. As a consequence, one would like to increase the number of features used in the discrimination process. Assume the balls are made out of a different material. Then the weight of the balls can be used as a discrimination feature. The ensemble membership of balls drawn from a box in a dark room would be established on the diameter and weight measured to determine where they would lie on a two-dimensional scatter diagram.
(Figure 6) It should be noted that the locus of points which are an equal number of black ball standard deviations from the black ball mean and an equal number of white ball standard deviations from the white ball data ensemble mean determine an equalilikelihood line. To the right of this line, the nearest mean is the black ball ensemble and any ball drawn which, by its weight and diameter, positions the point to the right of the mean is most likely a black ball. Balls drawn to the left of the mean are most likely balls which are members of the white ball ensemble.

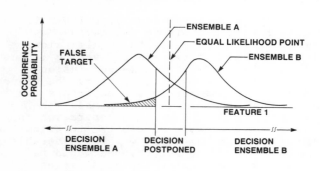

Figure 5

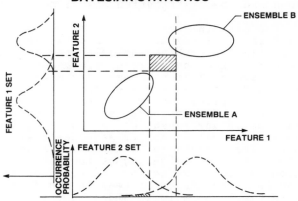

Figure 6

4. SENSOR FUSION APPROACH

The statistical approach in a single channel to discriminate a target presence involves a selection of features which separate the target from the background clutter by the greatest degree. Figure 7 describes such a three-dimensional feature space. Now the locus of points which are the same number of target standard deviations from the target distribution mean and background standard deviations from the background mean constituted equalilikelihood surface. Any point, measured in terms of the three features for the decision space, which lies to the right of the equalilikelihood surface is more probably a target than a background point. Any point, again measured in terms of the features of the decision space, to the left of the equalilikelihood surface is most likely a background point and not a target point. The designer immediately starts to expand the number of features used in defining the feature space to a larger and larger number in hopes of getting more authoritive decisions in determining the ability of the signal processing scheme to discriminate target from background. Unfortunately, this is a process of diminishing returns with the maximum number appropriate being approximately ten. In addition, the process is limited to some degree by the ability of the designer to either estimate, measure, or predict the statistical distribution in the ten-feature space.

Consider a dual mode infrared millimeter wave system using Quadratic Bayesian decision based target acquisition logics and weighted sensor fuzed schemes. (Figure 8) The millimeter wave system would measure each element in the background scene which is determined to be a candidate target in terms of the ten features used to define the millimeter wave decision space. The ten-feature space could include, for example, total radar cross section, minimum radar reflection center magnitude, the number of radar reflection centers separated by distances of a given magnitude, number of reflection centers having a magnitude greater than a clutter adaptive running average threshold, target width, edge estimators and others. (Figure 9) The infrared channel could use a different set of features involving the spatial separation of thermal sources in range and cross track, the magnitude of a thermal source, the total infrared signature and loss per stradian, the number of infrared sources on the target above a given clutter adaptive threshold, a number of infrared sources separated by a given distance and other concepts. (Figure 10)

QUADRADIC BAYESIAN DECISION SURFACE

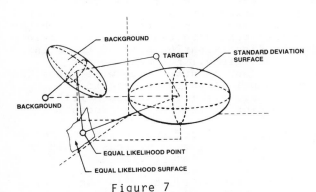

Figure 7

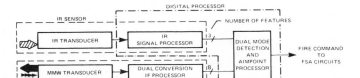

Figure 8

SPATIAL FREQUENCY ESTIMATE

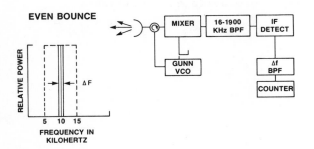

Figure 9

BASIC PRINCIPLE

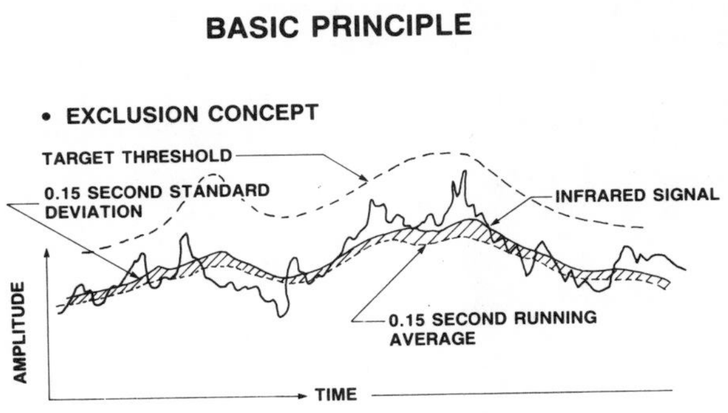

Figure 10

FEATURE SET SELECTION

MODIFIER DECISION SPACE

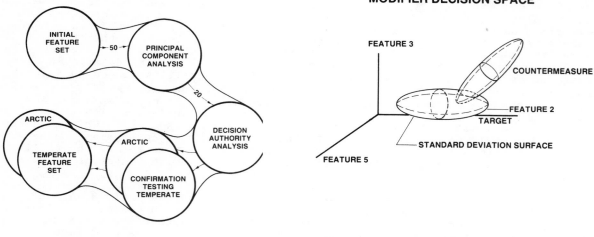

Figure 11

Figure 12

The selection of the features in the infrared channel and in the millimeter wave channel would be such to provide maximum separation between the statistical ensemble describing the target in the feature space and the statistical ensemble describing the background clutter.

The features selected to provide maximum separation between the target and background statistical ensembles in a temperate climate are not necessarily the same as the features to provide maximum separation between the target and background statistical ensembles in an arctic environment are not necessarily the same as those associated with a decision process tailored to equatorial or desert climates. (Figure 11) This requires that the process which identifies the feature space for the discrimination of the target from competitive returns from the background include measurements of both the target and background signature, employing the proposed candidate features, in a variety of backgrounds. Some of the features utilized in the discrimination process for either the infrared or millimeter wave channel will be identical in all features spaces employed for arctic, temperate, desert and equatorial measurements. Some will not. The features also, in general, cannot be expected to be identical to those used to discriminate the target in the presence of weather and diurdinal variations. All of these variations need to be accounted for in the selection of the feature process.

Perhaps the most interesting of all the problems in the sensor fusion schemes for a dual mode sensor lies in the area of the greatest role for dual mode systems. This is countermeasures. While it is easily argued that dual mode systems provide greater countermeasure discrimination capability than single mode systems, the exact implementation of a dual mode system to provide this discrimination has, in the past, focused only on the most simple schemes. These schemes have used simple and/or logics in the discrimination of countermeasures and have focused primarily on signature reduction or false target decoy schemes. It can be argued that if the enemy design process is effected correctly, the enemy will attempt to provide countermeasure

DECISION SPACE

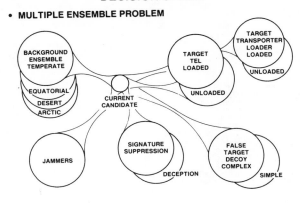

Figure 13

QUADRADIC BAYESIAN DECISION SURFACE

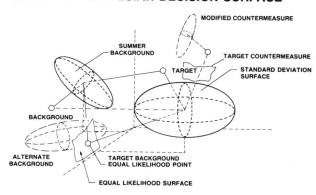

Figure 14

capability which closely emulates the target, in false target decoy schemes, or closely emulates the background in signature suppression or obscuration schemes. In either case, it can be argued that the ability to discriminate countermeasures becomes important in the sensor fusion process. Figure 12 details a three-feature decision space describing a false target decoy and a target ensemble. It should be noted that the false target decoy, if constructed, even in three or more feature spaces, will share a common intersection of the statistical distribution between target and countermeasure. The simplest way to look upon this is that those features used to discriminate the target from the background clutter, as previously discussed, are not necessarily the features which one should use to discriminate the target from the countermeasure. The ten-dimensional feature space used to effect this discrimination requires a different set of features in general and a different set of measurements. As a consequence, in the sensor fusion process there would be a modified decision space or scratch tab space in which a decision would be made on whether or not the current candidate being considered as a target, in the target acquisition logic which primarily separates the target from the background clutter, is a countermeasure. This would be affected in a ten-dimensional feature space or scratch tab space which would use ten different features to compare the target with a known countermeasure ensemble. The nearest mean calculation would be effected to determine whether or not it was more probable that a candidate was indeed a target or countermeasure.

In the final analysis, the decision space which one has to deal with in sensor fusion of something as simple as a dual mode system involves, if a statistical approach is employed, a multiplicity of features. Ten features would be used to discriminate the target from temperate background clutter. Additional ten would be used to discriminate the target from equatorial background clutter. Ten features each would be used for desert and for arctic clutter. Variations in the target construction typically have to be included. If the target is a transporter-erector-launcher loaded, it would have a different set of features than a target transporter-erector launcher unloaded. If false target decoys are present, whether they are simple or complex, a set of features has to be defined to separate them from the target statistical distributions. (Figure 13)

In general, the decision space, again reverting to three dimensions becomes extremely complicated since the nearest mean calculations have to be effected using the correct target and background data ensembles, as measured in the select feature space, to determine whether or not the current possible target candidate lies nearer to the target than to the background mean. In addition, scatch tab calculations have to be effected to determine whether the current candidate target, in the scratch tab feature space lies nearer to the target mean or the countermeasure mean. Several such countermeasure feature spaces may have to be established. (Figure 14)

DECISION LOGIC

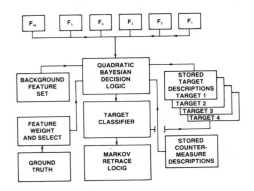

Figure 15

The sensor fusion discrimination logic then employs the features as measured in each of the two channels to effect (Figure 15) nearest mean determinations. These nearest mean determinations are against stored target descriptions which may include target descriptions which differ one from the other for the same target to accommodate the difference in the feature space when the target is approached from the head, tail, beam, or corner aspect. In addition, stored countermeasure descriptions will be provided in the Quadratic Bayesian decision feature space characterizing the discrimination of targets from countermeasures. The background feature set would also include arctic, temperate, equatorial, and desert descriptions. The ground truth input in terms of feature weighting and selection would be provided to the munitions prior to delivery against the targets to allow the munition to preset weights. It is important the munition be set for use in the winter or the summer. It is important that the munition logic knows whether or not its raining. It may even be important that the munition know essentially the time of day, to a reasonable interval, at which the attack will take place. This allows the decision process to select the appropriate features and weight the features in the voting between the modes of the multimode sensor.

5. SUMMARY

The problem appears complicated. It is. It is even more complicated for the munition designer since, as previously mentioned, the munition designer does not have the data in terms of target signatures for most of the high value or mobile force targets which he is designing munitions to engage and defeat. One only knows what they know. They know no more than that. As a consequence, the statistical approach offers the best method for the decision process and sensor fusion schemes for precision guided and sensor fuzed munitions. This is because, by its very nature, that it accommodates errors in the distribution. If properly chosen, where the feature space selected will separate to the maximum degree possible, the statistical distributions associated with the target and background clutter or target and countermeasure, for the given set of environmental and climactic conditions allowing the nearest mean process to be very authoritive. It does not require that the standard distributions associated with the target in terms of the ten features be known exactly. They only need to be known generally since the nearest mean calculation, with up to ten features, has been shown to fairly authoritive. It is only in those cases where the statistical distribution for the background and the target are in immediate proximity in the n-dimensional feature space that the decision process becomes flawed to a certain degree. In these cases it is necessary that the measurements be more exact. The designer, however, in the sensor fusion scheme tends to select features in the definition of the feature space which prohibits these sort of overlaps. The logic detailed here describes significant voids and deficiencies in the current developmental programs. Certainly the measurement efforts should include measurements in arctic, temperate, equatorial and desert backgrounds. Measurements in training programs should consider diurdinal and weather variations. But most importantly, perhaps, the effort should include consideration of countermeasures. The countermeasure communities should work closely with the munition developers not to indentify weaknesses in the developmental programs -- causing their termination for perceived inadequacies in performance -- but in the sense of helping the munition designer accommodate countermeasures in the sensor fusion process. Countermeasures against precision guided and sensor fuzed munitions used to engage mobile targets should be included in the test and development process. It should be recognized that the enemy is willing to spend considerably more resources in both manpower and material in protecting certain high value fixed targets. Countermeasures of these types should also be included in the test and evaluation process. The task of the enemy countermeasure expert is not an easy one either since he has to accommodate climactic and weather variations in the definition of countermeasures. False target decoys to be employed in a temperate environment, in general, must be different than false target decoys employed in an arctic environment. All of these features need to be considered in any signal processing or sensor fusion scheme which will work well most of the time.

The development community clearly needs to understand what it knows and what it does not know. In terms of the multi-dimensional decision process described, it knows little about arctic, equatorial or desert climates. It knows little about countermeasures. It knows little about the signatures of advanced enemy land combat vehicles and certain enemy high value fixed target signatures. The use of mathematics which far exceed the knowledge of the target data base, climate, and diurdinal variations, and countermeasures through the use of complex and sophisticated mathematics is not appropriate. It is akin to solving a problem to ten decimal point accuracy while significant figure analysis suggests the answer can be determined only to one or two significant figures.

A Distributed Sensor Architecture for Advanced Aerospace Systems

Jeff Schoess and Glen Castore
Honeywell Systems and Research Center
Minneapolis, MN.

February 29, 1988

Abstract

The Distributed Sensor Architecture (DSA) has been developed to couple knowledge-based processing with integrated sensors technology to provide coherent and efficient treatment of information generated by multiple sensors. In this architecture multiple smart sensors are serviced by a knowledge-based sensor supervisor to process sensor-related data as an integrated sensor group. Multiple sensor groups can be combined to form a reconfigurable, fault tolerant sensor fusion framework. The role and topology of this architecture are discussed. An example application of DSA sensor data fusion is presented.

1 Introduction

The availability and the need for multisensor systems is growing, as is the complexity of these systems in terms of the types and numbers of sensors within a system. The need to make simultaneous real-time measurements of multiple physical parameters is essential in advanced space-based structures and hypervelocity vehicles due to the complex structural dynamics, hypersonic flow and supersonic combustion processes. Most aerospace sensor-based systems to date have been designed around a small number of analog-based sensors and integrated with existing centralized processing resources. These systems typically exhibit severe computation, data bandwidth, and fault tolerance limitations as sensing and computational needs increase. In the future, however, such systems will need to operate in reconfigurable multisensor environments; for example, next-generation rocket propulsion systems for the National Aerospace Plane (NASP) and unmanned automated cargo vehicles will incorporate embedded propulsion and health-management subsystems to assess and diagnose engine faults. Another area of application currently receiving attention is manufacturing automation. The ability to provide sophisticated sensing and control capabilities in robotics, machining, and process control applications is crucial to increasing productivity through advanced automation.

There are at least four issues central to the design and effective deployment of multisensor systems:

1. *specification* of sensor characteristics and properties in terms of physical laws and uncertain sensor geometry

2. *development* of methodology for sensor data analysis and compensation techniques

3. *organization* of sensors to fulfill a specific roles

4. *management* of sensors in terms of providing greater tolerance for sensing device failure, and generation of appropriate control tasks.

The Logical Sensor Specification (LSS) [1] is a methodology for specifying multisensor systems and their implementations. The purpose of the LSS is to provide an implementation independent description of a sensor which identifies sensor characteristics and sensor compensation techniques. Logical sensors are a means by which to insulate the user from the pecularities of particular physical sensors.

In a complementary manner, the Distributed Sensor Architecture provides the framework for transforming the logical representation of a multisensor system described by LSS into a hardware implementation which has the capability of interpreting and fuseing data provided by multiple sensors, and controlling sensor activity. The DSA is an efficient mechanism for the aquisition and interpretation of data provided by multiple physical sensors.

The DSA concepts discussed in this paper have been developed as an outgrowth of related research projects. These projects include the implementation of the Space Station Integrated Sensor and the development of a software architecture, the Real-Time Blackboard Architecture (RTBA), for building embedded expert systems.

2   Logical Sensor Specification

A logical sensor specification describes a distributed sensor system as a data flow network of logical sensors. The purpose of such a specification is to provide a reasonably complete and structured description of a sensor network which can then be used for model-based control or expert systems applications.

A logical sensor, as shown in figure 1, has four primary components:

1. A *logical sensor name*. This is used to uniquely identify the logical sensor.

2. A *characteristic output vector*. This is a vector of types, such as boolean or integer, which serves as a description of the output of the logical sensor. The type may be any standard type (e.g., real integer), a user-generated type, or a well-defined subrange of either.

3. *Alternate subnets*. This is a list of one or more alternate ways in which to obtain data with the same characteristic output vector.

4. A *selector* whose inputs are alternate subnets and an acceptance test name. The role of the selector is to detect failure of an alternate and switch to a different alternate.

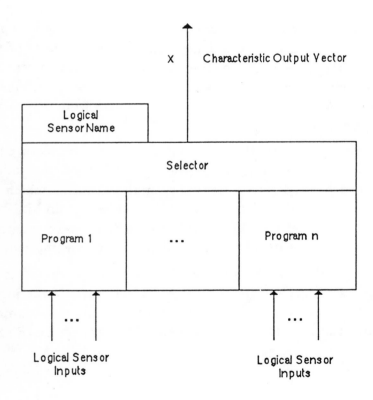

Figure 1: Graphical View of a Logical Sensor

A logical sensor can be viewed as a network composed of subnetworks which are themselves logical sensors. Communication within a network is controlled via the flow of data from one subnetwork to another. Hence, such networks are *data flow* networks. There may be alternate input paths to a particular logical sensor, and these correspond to the alternate subnets.

Each alternate subnet associated with a given logical sensor is equivalent, with regard to type, to all other alternate subnets in the list for that sensor and can serve as a backup in case of failure. The program of an alternate subnet accepts input from the source logical sensors, performs some computation on them, and returns as output a set (stream) of vectors of the type defined by the characteristic output vector.

3 Distributed Sensor Architecture

The Distributed Sensor Architecture (DSA) System is proposed as an efficient and logical approach for alleviating many limitations associated with centralized architectures. The control and sensing strategy implemented by DSA is based upon partitioning control responsibility into manageable, distributed units which perform significant numeric symbolic processing locally, within an integrated sensor. The DSA is organized hierarchically in three levels: *integrated sensors*, *sensor supervisors*, and *supervisory communication and control*. At the lowest level, the integrated sensor (IS) incorporates several analog transducer elements, signal conditioning electronics and a microcontroller integrated together into one or two IC chips. A sensor supervisor (ISS) communicates with one or more integrated sensors over a serial sensor communications bus. The ISS provides supervisory assistance to each IS including: sensor data interpretation and analysis, data trending, control command generation, and sensor to sensor association. The DSA topology is illustrated in figure 2.

Sensor processing is distributed within the multisensor system between several ISS's. Each ISS is allocated a set of integrated sensors based upon system topology or functional requirements. Sensor supervisors communicate using a shared memory accessed through a parallel high speed bus. Communication between the DSA and other subsystems occurs over a local area network (LAN) through the Network Interface Unit (NIU).

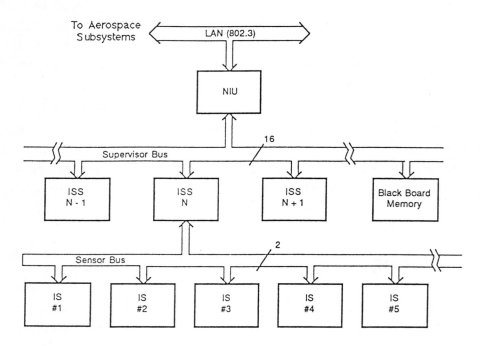

Figure 2: Distributed Sensor System Topology

To support symbolic processing the DSA implements a distributed blackboard architecture which facilitates sensor data fusion within a single sensor group (controlled by an ISS), among sensor supervisors, and among subsystems (the structure of blackboard systems is discussed in the next section). This architecture employs two different types of blackboards, a sensor supervisor blackboard and system blackboard. The supervisor blackboard resides in the supervisor's memory space and supports sensor data fusion among integrated sensors allocated to the ISS. At a higher level, the system blackboard exists as a shared memory accessed through the parallel supervisor bus. Supervisors can communicate through this blackboard to fuse data among interactive sensor groups. Sensor-related reports located on the blackboard would be passed as messages to other subsystems through the NIU to facilitate data fusion across the entire system

The DSA has been designed to support dynamic reconfiguration of the sensor system, to achieve a high degree of fault tolerance, to reduce complexity through the use of shared resources, and to allow real time monitoring of sensors.

System Reconfigurability: In terms of DSA hardware, integrated sensors and sensor supervisors can be configured to acquire and analyze sensory data, as a part of a dedicated aerospace subsystem, between two subsystems, or as a standalone mullti-sensor system: at the sensor level, integrated sensors can be reconfigured to perform on alternate function based on environmental conditions: for example, a pressure sensor could be reconfigured as a temperature sensor when given an appropriate control command from the sensor supervisor.

Fault Tolerance: The distributed topology of DSA suggests that faults at supervisor or sensor nodes will incur failure, but the system as a whole will continue operating. A sensor bus-driven architecture supports redundancy management principles where redundant integrated sensors can be activated by the sensor supervisor if the sensor's built-in-test (BIT) function indicates a sensor failure [2]. In addition, BIT capability can serve as aerospace subsystem diagnostic tool to isolate up to 97% of related hardware failures.

Shared Resources: This sensor bus-driven architecture promotes the idea of sharing sensor and control resources between integrated sensors, sensor supervisors, and aerospace subsystems; for example, an atmospheric pressure and oxygen controller in an environmental control and life support system could use ancilliary temperature data provided by the module temperature and humidity controller to control the module air pressure to within specified limits. Shared resources facilitate reducing overally system complexity by reducing wiring requirements (due to sensor bus utilization), power dissipation, system size and cost.

Real-Time Sensor Monitoring: A fully distributed sensor system permits sensory data to be acquired continuously to detect specified phenonema which may be random in nature (centralized sensor systems can only acquire data in a time-multiplexed fashion).

4   Description of DSA Elements

The topology of the DSA and the roles of each of the primary components have been described briefly in the previous section. In this section a more detailed description of the structure and function of each of these components is provided.

Integrated Sensor

An integrated sensor is device with one or more transducer elements, signal conditioning and signal processing electronics, microcontroller, and communication circuitry integrated into the same package [3]. These functions may be on the same IC chip or on separate chips. Some key features of the integrated sensor are:

1. a solid-state transducer array,

2. on-board signal amplification electronics,

3. A/D converter, sensor built-in-test electronics,

4. an 8-bit microcomputer with associated program and data memory,

5. sensor bus electronics.

The solid-state transducer array incorporates several thin-film based transducer elements into a multilayer array. These include piezo-resistive pressure transducers, nickel iron RTDs, microbridge air-flow transducers and metal Hall-effect devices [4]. In figure 3 the integrated sensor is shown functionally partitioned into two units: the analog transducer unit (ATU) and microcontroller unit (MCU). The ATU contains signal amplification and transducer conditioning electronics as well as a programmable-read-only memory (PROM) which contains the appropriate characterization coefficients for each transducer element in the transducer array.

The integrated sensor's embedded microcomputer provides several important features including: digital compensation (for sensor nonlinearity and slope/offset), data interpretation and analysis, data storage and formatting, time stamp referencing, and supervisor command interpretation and command handling. In addition, the microcomputer executes the sensor's built-in-test function on command or on a periodic basis to isolate hardware failures.

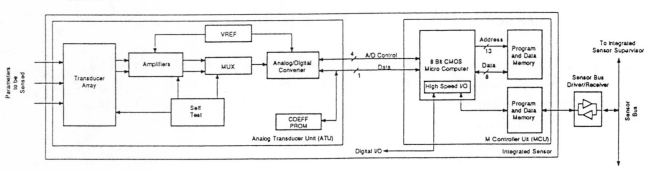

Figure 3: Honeywell Integrated Sensor

Integrated Sensor Supervisor

The second DSA element is the integrated sensor supervisor. The ISS is a knowledge-based controller which implements Honeywell's Real-Time Blackboard Architecture (RTBA) [5]. The RTBA is a general software architecture which serves as a base for building real-time embedded knowledge-base systems. The architecture framework is a form of a blackboard system concept whose details are discussed elsewhere [6]. A blackboard system can be thought of as consisting of four types of elements:

1. *Entries:* These are sensor-related inputs and intermediate results (i.e. interim goals) generated during problem solving.

2. *Knowledge Sources:* Independent, event-driver processes which contain a dedicated set of sensor-specific rules which describe how sensor data should be identified and selected based upon a priori sensor physical properties, and used to solve a specific problem.

3. *Blackboard:* A structured, global data base that contains sensory knowledge, sensor rule execution results so far and provides communication among knowledge sources and the system blackboard.

4. *Knowledge Source Control Mechanism:* A CPU which decides if and when particular knowledge sources (and associated rules) should generate entries and record them on the blackboard.

All blackboard systems are structured according to this high-level description, but their designs differ considerably at more detailed levels. The implementation of RTBA within the ISS is shown in figure 4. RTBA differs from traditional blackboard systems by including a structured knowledge source rule language together with a framework for explicitly specifying the structure of the control mechanism that chooses the next knowledge source. RTBA uses forward-chaining rules that have if-then-else and case structures and are executed sequentially thereby eliminating the implicit control of "conflict resolution". Knowledge sources are selected using an explicit knowledge source control diagram, eliminating the implicit control of a "scheduler". These features make RTBA expecially appropriate for applications involving intelligent sensor supervision which execute actions in a manner consistent with the constraints of a real time system.

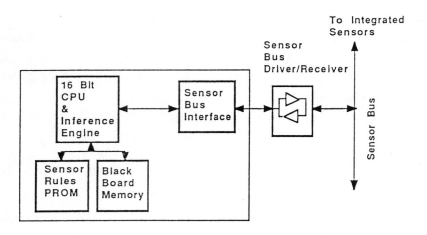

Figure 4: Honeywell Integrated Sensor Supervisor

Knowledge-based technology gives the ISS the ability to make intelligent system-related decisions based upon sensor physical laws, relationships between interactive sensors and sensor groups and mission requirements. These decisions could include:

1. *Data Prescreening:* ISS could evaluate usefulness of data at the time of acquisition by identifying the occurrence of pre-defined features in the sensory data.

2. *Data Selection:* Sensory data could be preselected to be recorded and analyzed off-line or to be used to generate appropriate control actions based upon sensor data content (i.e. identification of specific physical phenomena) or sensor time-related event.

3. *Control Command Generation:* Generate specific actuation commands for integrated sensor based upon interpretation of sensory data.

4. *Redundancy Management:* Poll and analyze built-in-test (BIT) data from each integrated sensor on a periodic basis and, if BIT analysis identifies hardware failure, replace faulty sensor with redundant sensor.

Sensor Bus

The third DSA element is the sensor bus which supports communications between each integrated sensor and ISS. The sensor bus is a 2-wire bidirectional serial bus utilizing a poll-response protocol. This means that the sensor bus master polls each slave for status or data by issuing a unique slave address. The designated slave recognizes its own address and responds by transmitting a slave acknowledge word followed by 1 to N data words. The DSA system defines the ISS to be sensor bus master and integrated sensors to be bus slaves. Our work has focused on low power implementations of sensor bus technology. We have designed and developed the Honeywell Control Point Bus (CPB) [7] for industrial and military control system applications. The CPB supports two types of polling schemes: direct polling and significant event polling. Direct polling is characterized by a master directly addressing a single slave to request data, send data or issue a command. Significant event polling features the ability to broadcast a message to a group of slaves or all slaves at once. If there are no replies, the master assumes that there has been no "significant event" since the last poll. However, if two or more slaves respond, the CPB resolves the bus collision by implementing a "class polling backoff algorithm." A typical example of a significant event would be: *the engine compressor is* 1500°C, *when normally it is* 1000°C.

5 An Example Application: Sensor Data Fusion

The following simple example demonstrates the power of the DSA approach for distributed sensor data fusion. The goal is to acquire sensory data from multiple sensors and fuse (logically combine) the resultant data to make a decision or to acquire knowledge of a situation.

In the DSA system sensor information is represented in a specific format. The sensor supervisor environment E is structured as an unordered set of integrated sensors, S_i, so that E = (S_1, S_2, ..., S_n). Each integrated sensor, S_i, in this set has an associated set of transducer elements, T_{ij}, or $S_i = (T_{i1}, T_{i2}, \ldots, T_{in})$. As previously described, these transducer elements output on electrical signal proportional to physical phenomena (pressure, temperature, etc.). Associated with each transducer element is a 4-tuple, (v, e, c, t_s) where v is single data value, e is an estimated error value, c is confidence value, and t_s is a time stamp.

In this expression the transducer data value, v, refers to the raw data reading converted to an equivalent digital word. The error value parameter, e, is an accuracy estimate based on the knowledge of the sensor environment and transducer transfer function characteristics. The confidence value, c, is a measure of the faith in the transducer model and the ability to interpret property values and error estimates. The confidence value can be used as a criterion to choose between conflicting sensor estimates that have the same properties. The timestamp parameter, t_s, can be used to fuse complementary sensor data which involves time-variant physical phenomena.

An example of how a distributed blackboard architecture can fuse sensor data is shown pictorially in figure 5. Integrated sensor supervisor blackboards A and B each contain a list of sensor-property tuples for respective integrated sensors 1 and 2. This sensor data is shown posted on the shared memory blackboard by the integrated sensor supervisors in the form of a sensor-related report. The fused data value V_f is then determined by solving a Boolean expression which compares the data values for transducers 1 and 2. If transducer 1 is greater than or equal to transducer 2 and has a higher confidence value, the value of V_f is equal to the data value for transducer 1. Otherwise V_f is equal to data value for transducer 2.

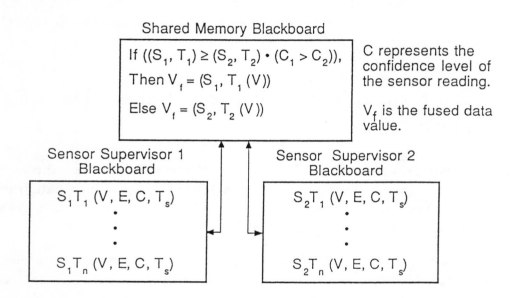

Figure 5: Example of Distributed Blackboard Data Fusion

This example illustrates how transducer confidence levels can be used to choose between conflicting sensor estimates. If, for example, the data value of transducer 1 is greater than transducer 2 and the transducer models are identical (i.e. measure the same physical phenomena) a conflict exists: em has transducer 1 failed or transducer 2? This conflict can be resolved by comparing the confidence values for each transducer estimate. In the example above, transducer 1 may have failed if its confidence value is less than the value of transducer 2. This simple example illustrates how sensor data fusion provides both fault detection and isolation capability and knowledge of the parameter being measured.

6   Summary

The DSA system represents an innovative and novel framework for performing sensor data fusion. The computing architecture discussed above facilitates data fusion at the sensor level, integrated sensor supervisor level, and the aerospace subsystem level through the distributed blackboard topology. The implementation of knowledge-based reasoning techniques in the ISS in concert with the incorporation of microcomputer-based intelligence at the sensor level together form the basis for a multi-sensor system.

References

[1] Tom Henderson and Ester Shilcrat, "Logical Sensor Systems" Journal of Robotic Systems, vol. 1, no. 2, 1984 (169-193).

[2] Earl Benser and Ken Kawai, **A Review of Integrated Sensor Technology**, Yamatak, Honeywell/Systems and Research Center Project, 1986. (26-31).

[3] S. Middelhock and A.C. Hoogerwerf, *IEEE Transducer '85 Digest*, 1985. (2-7).

[4] J.C. Anderson, "Thin Film Transducers and Sensors", *Journal of Vacuum Science Technology*, May/June 1986. (610-616).

[5] K. Whitebread, RTBA - "A Software Architecture for Real-time Embedded Expert Systems", Honeywell/SRC, 1986. (unpublished).

[6] Hayes-Roth, Barbara, "The Blackboard Architecture: A General Framework for Problem Solving: Heuristic Programming Project", Report No. HPP-83-30, Computer Science Department, Stanford Univ., May 1983.

[7] Sensor Bus Communications System, *CTRC Report*, June 1985.

SENSOR FUSION

Volume 931

Session 2

System Aspects

Chair
Charles B. Weaver
Honeywell Electro-Optics Division

Spectral analyzer and direction indicator (SADI) system

Jerome M. Welner

Hughes Aircraft Company, Sensor Systems Engineering Laboratory
2000 E. El Segundo Blvd., El Segundo, California 90245

ABSTRACT

The spectral analyzer and direction indicator (SADI) system is an electro-optic sensor that can determine both the spectral content and direction of the source of a single pulse of radiated energy. In its simplest form, it can determine the direction of the source in one dimension. Directional information in two dimensions requires the addition of some electronics and, in some configurations, optics. Various degrees of electronic processing sophistication may yield the pulse shape, and its spectral and spatial distribution.

1. INTRODUCTION

It is often desirable to have hardware that can answer one or more of the following questions:
Is there a laser being fired in the vicinity?
If there is a laser, what is its wavelength?
Is the source of radiation a laser or broadband?
Where is the source of radiation?

The SADI system was invented to provide hardware to answer these and other questions. The SADI is a staring system providing continuous coverage within its field-of-view. Within that field it will, as a minimum, locate a source of radiation and determine its wavelength or spectral content. Given the proper detector array and processing electronics, it will determine the temporal and spectral signatures of the source. Position may be determined in one or two axes. All these data may be determined from a single received pulse.

2. BASIC CONCEPT

The angle at which energy leaves a given diffraction grating is a function of the angle of incidence of the incoming radiation and its wavelength. The output angle may therefore be thought of as the solution to an equation with two unknowns. A typical grating spectrometer reduces the situation to an equation with one unknown by carefully controlling the angle of incidence. Wavelength may then be determined uniquely from the diffraction angle. Figure 1 shows a schematic representation of a spectrometer using a reflective grating.

The diffraction angle for a reflective grating is found from:

$$\sin \theta_D = \frac{N\lambda}{D} - \sin i \tag{1}$$

in which:

θ_D = the diffraction angle
N = the diffraction order
λ = the wavelength of the radiation
D = the grating line spacing (same units as λ)
i = the angle of incidence of the radiation

By controlling N, D, and i, the wavelength of incoming radiation may be determined by measuring the diffraction angle.

It is obvious that if the entrance slit and lens of the spectrometer of Figure 1 were eliminated and light were allowed to enter from a large angle, there would be no way to determine either the light's angle of incidence or its wavelength. If, however, the angle of incidence of the light could be determined by other means, its wavelength could be computed.

The SADI system accomplishes this goal by adding a plane mirror to the grating spectrometer. The angle of incidence of the radiation is determined from the reflection angle from the mirror. This is shown schematically in Figure 2.

In essence, we now have two equations and two unknowns. To the equation above, we have added:

$$\theta_R = i \tag{2}$$

in which:

θ_R = the angle of reflection

As an example of how the system would separate laser lines, the following was analyzed:

- An instrument with a ten degree field-of-view that provides for incidence angles of 45 to 55 degrees

- Two different gratings; one with 600 lines/mm and one with 1200 lines/mm

- Three laser lines; ruby at 0.6943 μm, Nd:Glass at 1.054 μm, and Nd:YAG at 1.0645 μm

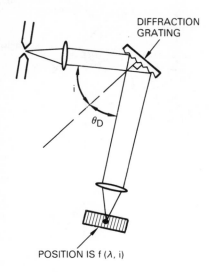

Figure 1.   Schematic representation of a spectrometer.

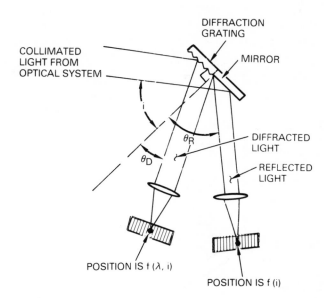

Figure 2.   Schematic representation of the SADI system.

The results of the analysis are in the following table.

Grating	600 lines/mm						1200 lines/mm					
i	45°			55°			45°			55°		
Laser*	R	G	Y	R	G	Y	R	G	Y	R	G	Y
θ_D	−16.889	−4.284	−3.922	−23.739	−10.763	−10.396	7.242	33.896	34.771	0.803	26.465	27.274
θ_R	−45°			−55°			−45°			−55°		

*R = Ruby, G = Nd:Glass, Y = Nd:YAG

Examination of these results yields some interesting information.

- The spacing between the Nd:Glass and the Nd:YAG lines is never closer than 0.36 degrees. This is readily detectable.
- For the 600 line/mm grating, there is a 21 degree gap between the largest θ_R (−45°) and smallest θ_D (−24°).
- The equivalent for the 1200 line/mm grating is approximately 46 degrees.
- For any single condition, θ_D and θ_R are never closer than 28 degrees for the 600 line/mm grating, or 52 degrees for the 1200 line/mm grating.

Figure 3 contains a graphic representation of the angular spread of light to and from the grating.

3.   IMPROVED CONFIGURATION

To be useful, the system must have some way to read the output angle. The obvious choice is a detector array in the focal plane of an optical system. From the example above it is seen that very wide angle optics and a large focal plane would be needed, particularly with the 1200 line/mm grating. An alternate choice would be two optical systems and detector arrays, one for the reflected light and one for the diffracted light.

However, there is a much better solution. It has been seen that the reflected light never gets closer than 28 degrees to the diffracted light when the 600 line/mm grating is used, or 52 degrees when the 1200 line/mm grating is used. If the mirror is tilted either 14 or 26 degrees with respect to the grating, depending upon which grating is used, these gaps will be almost closed. The reflected light still will always be on one side of the diffracted light and there will be no ambiguity in the results so long as the computations always adjust the angle of the reflected light by twice the mirror-grating offset angle.

The effect of tilting the mirror with respect to the grating is shown in Figure 4. The overall effect is a reduction in the required field-of-view of the output optics from 41 degrees to 23 degrees for the 600 line/mm grating system and from 90 degrees to 38 degrees for the 1200 line/mm grating system. If the output optics had a 100mm focal length, the detector array would be reduced in size from 75mm to 41mm for the 600 line/mm system and from 200mm to 69mm for the 1200 line/mm system.

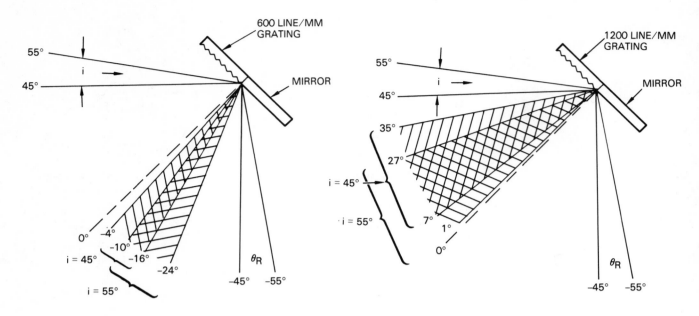

Figure 3. Response of two gratings and mirror to three wavelengths and two angles of incidence.

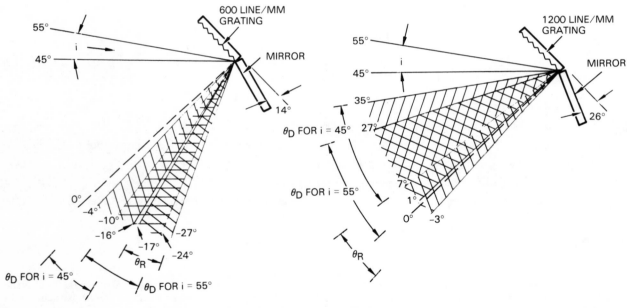

Figure 4. Effect of tilting mirror with respect to grating.

The diameter of the output optics must be large enough to encompass the extreme rays from both the mirror and grating. By increasing the offset of the mirror, the reflected light could be imaged on the long wavelength side of the diffracted light rather than the short wavelength side. This has negligible effect on the angular field-of-view required of the output optics but tends to fold the output beam over on itself, thereby reducing the required size of the entrance pupil. This modification is illustrated in Figure 5.

4. EXPERIMENTAL RESULTS

A very simple system was set up in a laboratory to illustrate the concepts of the mirror-grating system described above. As shown in Figure 6, two lasers were used providing three radiation lines: 0.4545 μm, 0.5145 μm, and 0.6328 μm. The grating had 600 lines/mm. The lens had a focal length of approximately 150mm. A mirror offset of approximately 14 degrees was used. This was sufficient to image the reflected light on the long wavelength side of the diffracted light.

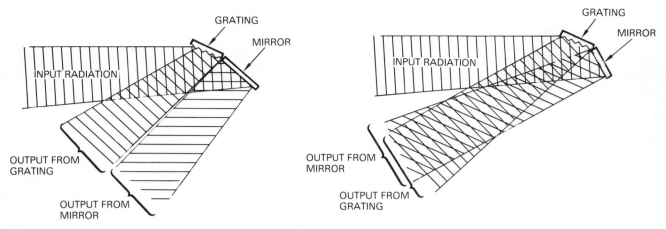

Figure 5.   Effect of increasing the mirror offset.

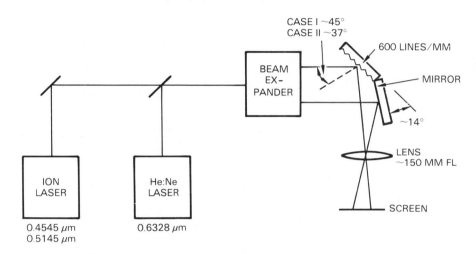

Figure 6.   Experimental setup.

Figure 7 contains two photographs of the experimental results. Each photograph was made with three exposures; one exposure was used for each color. Each exposure included two spots of the same color. The spot which became white was the superposition of three colored spots. These were reflected by the mirror and were superimposed because the incidence angle was held constant for each photograph.

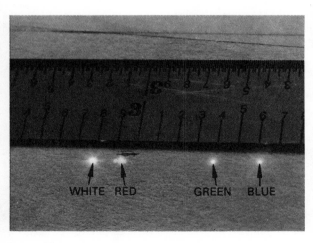

a. Incident angle ∼ 45°

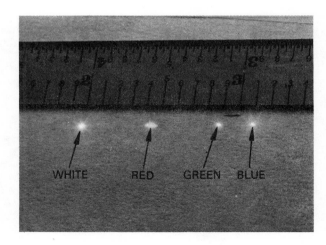

b. Incident angle ∼ 37°

Figure 7.   Photographs of experimental results.

5. OTHER MODIFICATIONS

Once the basic SADI idea is understood, various options become apparent. For instance, certain system requirements may allow for the elimination of the output optics by use of a concave mirror and grating. Also, some installations may have physical constraints that would be better met with a transmissive grating, or even a prism, rather than a reflective grating. Finally, for systems with limited bandwidth and field-of-view, holographic technology may eventually allow for the combination of the grating, mirror, and output optics into one holographic optical element.

6. TWO AXIS SYSTEM

As described so far, the system can determine the wavelength of a source and its angular position along one axis. However, information about the source position along the other axis is already present in the output plane. If we just consider the image of energy reflected by the mirror, it is obvious that it contains two-dimensional information that can be readily extracted.

Assume a system in which the grating grooves are vertical and the output is taken from a detector array at the focal plane of the output optics. Figure 8 shows the focal plane divided into three regions. In Region I only energy from the mirror may be imaged. Energy from either the mirror or the grating may be imaged in Region II. Only energy from the grating may be imaged in Region III. Regions I and II, therefore, contain the total angular output of the mirror, and Regions II and III contain the output of the grating.

From Figure 8b, it is obvious that two-dimensional information may be extracted from the system by using a two-dimensional detector array in regions I and II of the focal plane. Region III is used only for spectral information, therefore, a one-dimensional array may be used.

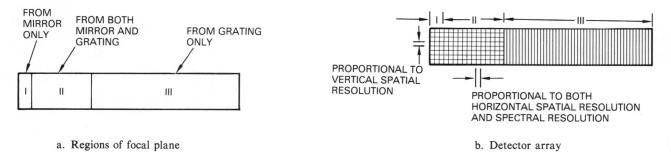

a. Regions of focal plane

b. Detector array

Figure 8. Focal plane and detector array for two-axis system.

7. PROBLEMS—REAL AND IMAGINED

Questions about the SADI system generally involve one of four subjects: simultaneous events, broadband radiation, extended sources, and system sensitivity. These are discussed briefly below.

7.1 Simultaneous events

Simultaneous events may be caused by the simultaneous firing of more than one laser or by energy from one laser being scattered by more than one object. Given typical laser pulse widths and the fact that the detectors must receive the energy simultaneously for a detection to occur, it is very unlikely that two independent lasers would be detected in the same field-of-view at the same time. Therefore, what the system would most likely see would be two separated sources having the same wavelength.

The system would see four spots, two from the mirror, and two from the grating. The spot at one end will always be reflected from the mirror. This defines the location of one source. As this spot is paired with each of the other three spots, the two remaining spots are paired with each other, and three pairs of wavelengths are defined. If the spot from the mirror is called #1, the possibilities of position and wavelength are shown in the table below.

Positions defined by:	Wavelengths defined by:
1 & 2	1,3 & 2,4
1 & 2	1,4 & 2,3
1 & 3	1,2 & 3,4

In the real world it would be almost certain that one pair of wavelengths would be identical, indicating that a single laser and its reflection were seen. Because timing may be held to nanoseconds, the probability of seeing two independent events in the same field-of-view may be made vanishingly small. However, if two lasers were seen at the same time, it is likely that one pair would contain at least one known laser line. This should be selected as the correct solution.

If one laser happened to put out two lines, the system would detect three spots. The spot from the mirror would always be known, therefore the other two would be identified as diffracted and would define two wavelengths.

It is possible, at least theoretically, to fool the system. In the unlikely event that two variable wavelength lasers happened to fire simultaneously in the field-of-view of the SADI system, the system would be confronted with a truly ambiguous situation and would output three possible solutions.

7.2 Broadband radiation

A source of broadband radiation, reflected by the mirror, would be imaged in the focal plane as a point; from the grating it would be imaged as a line. The point would still indicate the location of the source. The line characteristics would be indicative of the source's spectral characteristics. If the detection scheme of the instrument allowed for the measurement of amplitude, the spectral characteristics of the source could be determined.

7.3 Extended sources

The reflected image of an extended source will cover a discrete area in the focal plane. If the source is monochromatic, its diffracted image will cover a similar area. The processing electronics may still analyze the source by making this determination.

It is possible to see an extended source that radiates over a broad spectral band. To determine both the source shape and spectrum, it would be necessary to measure amplitudes accurately and deconvolve the diffracted image. Though theoretically possible, it might be beyond the capabilities of a practical instrument.

7.4 System sensitivity

There are three factors that reduce the sensitivity of the SADI system with respect to a "straight through" optical system: grating efficiency, polarization effects, and image splitting. Grating efficiency and change of efficiency with polarization can be optimized by proper blazing of the grating's grooves and orientation of the grating with respect to the incidence and diffraction angles. Though the effects can be minimized, they cannot be eliminated.

The division of the energy between two output spots is significant from a sensitivity standpoint. It is also the key to the SADI concept and provides the capability for achieving a false alarm rate that would otherwise be unachievable. This is discussed more fully below.

8. COINCIDENCE DETECTION

The SADI concept requires that events be detected in two places simultaneously. This coincidence requirement enables the regaining of some of the sensitivity lost by splitting the beam. If it is assumed that the noise in both channels that detect a coincidence is random and uncorrelated, it is possible to work at lower signal to noise ratios than otherwise, because a false alarm would require the noise in both channels to increase simultaneously beyond some threshold. Depending on the analysis bandwidth and the duration of the window that defines the acceptable time discrepancy for two signals to be considered as simultaneous, much of the lost sensitivity may be regained.

The primary advantage of coincidence detection, however, is not related to sensitivity. No manipulation of bandwidth and window time will allow the system to regain all the sensitivity lost in the energy split. The big advantage is in the reduction of false alarms.

One of the main features of the SADI system is that it allows single pulse detection of lasers in real time. If detection were indicated by a single detection channel exceeding a predefined threshold, any exceedance of this threshold, in any channel, would be a false alarm. If the only noise sources were background noise and system internal noise, the noise could probably be approximated by a convenient amplitude distribution, such as Gaussian or the sum of Gaussians, and have a monotonic decrease in the probability of events larger than one or two times the rms noise level. This would enable the achievement of any desired false alarm rate by setting the threshold of detection high enough.

To this well behaved noise, however, must be added very large pulses caused by cosmic rays or their products, such as Compton electrons. These pulses will have temporal characteristics approximating the impulse response of the detection system and amplitudes varying from the noise level of the system to system saturation, depending on the energy and charge of the particles causing them. In short, there would be virtually no way to separate the effects of cosmic rays from very short laser pulses in a system that requires detection in only a single channel.

Simultaneous detection in two nonadjacent channels precludes false alarms from single cosmic rays. False alarms from two independent cosmic rays would also be very rare. As stated previously, if the time window is made small enough, the probability of two independent events occurring within the window may be made very small.

Therefore, the apparent liability of the energy split required by the SADI concept is really a major advantage in the reduction of false alarms brought about by the use of coincidence detection.

9. CONCLUSIONS

The SADI concept may be used in a simple instrument to detect lasers. By increasing the system and processing complexity in small steps, the instrument may be made to:

- locate sources in one axis
- locate sources in two axes
- identify lasers by wavelength
- determine the temporal characteristics of sources
- determine the spatial, temporal, and spectral characteristics of arbitrary sources
- identify single sources with multiple wavelengths
- identify multiple sources with multiple wavelengths

Proof-of-concept model of a multichannel off-axis passive bidirectional fiber optic rotary joint

Walter W. Koch and Sam K. Nauman

KDI Electro-Tec Corporation

1600 N. Main Street, Blacksburg, Virginia 24060

Abstract

A multichannel off-axis passive bidirectional fiber optic rotary joint has been built with a derotating transmissive intermediate optical component (IOC) comprised of a nearly coherent optical fiber bundle with lensed transmitters and receivers. This proof-of-concept model has been tested for insertion loss and rotational variation. Insertion loss measurements vary from 3.2 to 5.9 dB and 4.0 to 9.9 dB at 1300 and 1550 nm, respectively. The variation in insertion loss values is due to several broken fiber paths inside the IOC. Improvements in manufacturing and packaging the IOC will allow the performance to approach the design goal of 3 to 5 dB insertion loss with less than 0.5 dB rotational variation at wavelengths of 600-1550 nm. Multichannel, off-axis, passive, and bidirectional light transmission capabilities over a broad range of wavelengths are important considerations to the system designer who must transmit multiple fiber optic remote sensing or communication channels across a rotary interface.

1.0 INTRODUCTION

With the increasing replacement of electrical data transmission systems with optical systems, many optical components functionally equivalent to electrical ones will have to be developed. Upgrading electrical data transmission circuits to fiber optic channels will require the use of a fiber optic rotary joint (slip ring). Speer and Koch[1] have reviewed the wide variety of fiber optic rotary joints and list the features which would be found in the ideal fiber optic rotary joint.

The ideal fiber optic rotary joint is capable of passively transmitting lightwaves through a number of channels in either direction off the axis of rotation. Each fiber optic channel should also transmit light with low loss and equal efficiency, regardless of operating wavelength and rotation angle. Multichannel capability will reduce or eliminate the need for multiplexing optics or electronics while providing channel redundancy. Passive light transmission through the rotary joint will allow active transmitters and receivers (LEDs or laser diodes and photodetectors) to be located remotely and protected from the potentially harsh environment of the rotary interface. Bidirectional light transmission, transmitting light with equal efficiency from stator to rotor and rotor to stator, will require a reciprocal optical design. The off-axis configuration will allow a pneumatic, hydraulic, or microwave rotary joint to occupy the axis of rotation giving a hybrid rotary joint package.

2.0 THE OPTICAL DESIGN

Optical continuity between rotor and stator channels is maintained by the use of a separate opto-mechanical device known as the intermediate optical component (IOC). The IOC performs optical derotation between rotor and stator through a coherent optical fiber bundle assembly geared to one half the rotation rate of the rotor.

Lensed optical transmitters and receivers are housed in the rotor and stator faceplates. This proof of concept design uses graded refractive index (GRIN) lenses to expand and collimate the light as it travels between the rotor and stator and vice versa. Alternately, laser or LED sources and photodetectors may be mounted directly on the rotor and stator faceplates in an active configuration.

2.1 Optical Derotation

The most elegant optical rotary joints incorporate either a transmissive or a reflective intermediate optical component which rotates at half the angular velocity of the rotor with the aid of a gear system. Optical continuity is determined by the combination of the IOC moving at half the angular speed of the rotor and the exact light path through the IOC.

Iverson[2] first described a rotary joint of this type using a Dove prism as the IOC. Koene[3] has commented on several disadvantages of derotating prisms in optical scanners and rotary joints and suggested a design using a glass sphere bisected by a thin mirror as the IOC to overcome these disadvantages.

Each of these IOC designs described so far must occupy the axis of rotation to maintain optical continuity properly between rotor and stator. Replacing a Dove prism with a carefully arranged bundle of optical fibers, as Birch[4] has suggested, eliminates the requirement that the IOC occupy the axis of rotation.

2.2 The Coherent Optical Fiber Bundle

Ideally, a rotary joint should transmit a coherent image. In theory, this is a simple task with a single channel on-axis design, but is significantly more difficult with a multichannel off-axis configuration.

The intermediate optical component of the proof-of-concept device is comprised of several hundred thousand specialized small diameter optical fibers placed in a symmetrically ordered array around a mandrel or cylindrical shell, as illustrated by Figure 1. Alternately, a similar design could be built with a platter shaped configuration. This careful arrangement of fiber segments ensures that the output faces are in the same relative position as the input faces.

Cylindrical shell configuration of Intermediate Optical Component

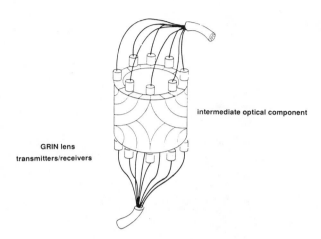

intermediate optical component

GRIN lens
transmitters/receivers

Figure 1. Cylindrical shell configuration of the intermediate optical component.

The effective numerical aperture of less than 0.15 will require the use of highly collimated light to enter the IOC. The proof-of-concept model uses transmitting and receiving fiber optic cables terminated with one-quarter pitch GRIN lenses to expand and collimate the light leaving each transmitting fiber, and to collect and focus the light into each receiving fiber.

2.3 Expanded Beams

Two fiber optic cables, each with one end terminated into graded refractive index (GRIN) lenses, are used for each optical channel. The function of the GRIN lens is to expand and collimate the beam of light as it leaves the transmitting fiber and travels across the air gap and into the IOC. As the expanded light beam exits the IOC and travels across the second air gap, it is collected by the receiving lens and focused into the receiving fiber.

The use of expanded beams will allow an increased tolerance to mechanical vibration, thermal expansion, and misalignment of the optical components due to shock or vibration. The expanded beam concept will also support any size or type (single or multimode, step or graded index) fiber optic cable used to pass light through the fiber optic rotary joint. Expanded beams will also permit using any combination of fiber optic cables and allow interchanging any number of channels without restricting future upgrades. The remaining end of each fiber optic cable may be terminated with any connector or splice to match the system requirements.

3.0 THE PROOF-OF-CONCEPT MODEL

The proof-of-concept model is a ten channel off-axis passive bidirectional fiber optic rotary joint. Figure 2 illustrates the proof-of-concept model, designated Electro-Tec part number 67810.

The design goals for the proof-of-concept model, part number 67810, were 5-8 dB insertion loss across a wavelength band of 600-1550 nm. With improvements in manufacturing the IOC, the design goal for insertion loss of the second generation of Electro-Tec part number 67810 will be 3-5 dB with rotational variation of less than 0.5 dB.

3.1 Insertion loss

Insertion loss measurements were taken at 1300 and 1550 nanometers with an Intelco 112 laser diode source and United Detector Technology S390 optical power meter with a model 261 germanium detector head. Figure 3 shows the test set-up. Sixteen readings were taken, one reading per 22.5 degrees of rotation. Table 1 summarizes the results. Polar plots of loss [dB] versus rotation angle are shown in Figure 4 for 1300 and 1550 nm.

Table 1.

| | | Insertion loss (dB) at | |
Sample	Rotor Position (°)	1300 nm	1550 nm
1	0.0	5.9	4.5
2	22.5	3.2	5.1
3	45.0	3.4	4.2
4	67.5	3.3	4.3
5	90.0	3.7	5.2
6	112.5	4.0	6.3
7	135.0	3.2	4.0
8	157.5	3.6	4.0
9	180.0	4.4	5.0
10	202.5	3.2	4.4
11	225.0	3.9	4.7
12	247.5	4.0	5.3
13	270.0	3.3	4.8
14	292.5	4.2	6.5
15	315.0	3.3	4.2
16	337.5	4.0	9.9
Mean Insertion loss		3.8 dB	5.2 dB

3.2 Rotational Variation

The variation in insertion loss is caused by imperfections in the IOC, as illustrated by the photomicrographs of Figure 5. Figure 5 (b) shows a joint between two neighboring fiber bundle segments where the IOC can no longer be considered coherent in the regions of broken fibers, kink bands, and undulations. The dark regions represent broken fiber paths where no light is transmitted. Improved fiber bundle segment-to-segment alignment and polishing and grinding techniques for the next generation of devices should reduce or eliminate these "dead" transmission zones. These improvements should allow the rotational variation to approach the design goal of less than 0.5 dB.

Proof-of-Concept Model

Electro-Tec P/N 67810

Figure 2. The proof-of-concept model designated
Electro-Tec part number 67810.

4.0 SENSOR FUSION CONSIDERATIONS

4.1 Multichannel Capability

Electro-Tec's design is able to support at least ten independent optical channels. The multichannel capability of the rotary joint will allow direct channel-for-channel replacement of conventional electrical data transmission circuits, reducing the need for multiplexing optics and electronics. Multiple channels will also allow for channel redundancy and back-up. As mentioned earlier in Section 2.3, this design will support any size optical fiber and permit "swapping" cables between channels. Spare channels may also be included in the design to allow easy system upgrade or reconfiguration.

Because the IOC is a continuous transmissive optical device, the number of channels is limited only by the number of GRIN lens transmitters and receivers that can be placed around the rotor and stator faceplates.

4.2 Off-Axis and Bidirectional Signal Transmission

It is often necessary for the system designer is to combine several rotary joints in a single package in a limited amount of space. This requirement gives the off-axis capability of the fiber optic rotary joint added importance. Most fiber optic rotary joints are included with conventional electrical slip rings and hydraulic, pneumatic, or microwave rotary joints to form a hybrid rotary joint package.

Bidirectional data transmission is a characteristic taken for granted in electrical systems. For optical systems, it requires a reciprocal optical design to pass light in either direction with equal efficiency. The coherent bundle of the IOC is a reciprocal device which will allow the bidirectional transmission of light signals from rotor to stator and stator to rotor.

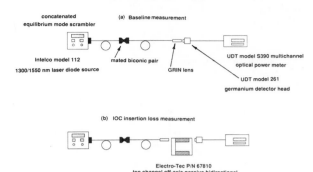

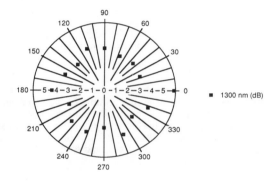

Figure 3. Insertion loss measurement test set-up
with (a) baseline reference measurement and
(b) IOC loss measurement.

Polar Plots of Loss vs. Angular Position
Proof-of-Concept Model

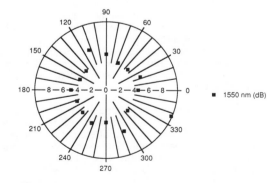

(a) Polar plot of loss [dB at 1300 nm] vs. angular position

(b) Polar plot of loss [dB at 1550 nm] vs. angular position

Figure 4. Polar plots of IOC insertion loss vs.
angular position.

Photomicrographs of Intermediate Optical Component

(a) region of low loss

(b) segment-to-segment joint with relatively high loss

Figure 5. Photomicrographs of IOC endface where (a) shows a region of low loss and (b) shows a segment-to-segment joint with relatively high loss.

4.3 Passive Light Transmission

This design will support passive as well as active light transmission through the rotary joint. Active light transmission is defined by the use of active opto-electronic devices (LEDs, laser diodes and photodetectors) to transmit and receive light signals across the rotary interface. The potential for a harsh or electrically noisy environment near the rotary joint may preclude placing active transmitters and receivers on the rotary joint. Passive light transmission allows the laser diode transmitters and photodetector receivers to be located remotely in a shielded and shockproof package, protecting the devices and maintaining signal integrity.

5.0 CONCLUSIONS

A ten channel off-axis passive bidirectional fiber optic rotary joint has been built and demonstrated successfully as Electro-Tec part number 67810. Insertion loss measurements vary from 3.2 to 5.9 dB at 1300 nm and 4.0 to 9.9 dB at 1550 nm. Improvements in the segment-to-segment alignment and endface polishing of the intermediate optical component should reduce the insertion loss to 3-5 dB and the rotational variation to less than 0.5 dB.

6.0 ACKNOWLEDGEMENTS

The authors would like to thank Alex Speer, Bill Jennings, and Ben Williams for their help in designing, building, and testing ETC part number 67810. We would also like to thank Galileo Electro-Optics (Sturbridge, MA) for their assistance in building the intermediate optical component.

7.0 REFERENCES

1. J. A. Speer and W. W. Koch, "The diversity of fiber optic rotary connectors (slip rings)," Society of Photo-Optical Instrumentation Engineers Volume 839, 1987.
2. M. L. Iverson, "Optical slip rings," U. S. Patent 4,109,998 (1978).
3. J. W. Koene, "Optical slip rings," U. S. Patent 4,447,114 (1984).
4. E. H. Birch and A. Labato, "Optical slip ring," U.S. Patent 4,460,242 (1984).

Adaptive control of multisensor systems

Su-shing Chen

Department of Computer Science, University of North Carolina-Charlotte
Charlotte, NC 28223

Abstract

A hierarchical and adaptive control scheme of multisensor systems is introduced for improve -ment of image understanding, correspondence (registration) problem and sensory data fusion. The neural network approach provides adaptiveness and learning to not only the control level of the overall system architecture, but also the processing level of the image frames. Furthermore, our improved sensing capability enhances the performance of large, complex, integrated sensor-driven robotic systems.

1. Introduction

In applications of large, complex, integrated robotic systems, sensory feedback control is an essential component. In [1],[2], a single spherical data structure for multisensor systems and control guidance systems is proposed. The (θ,ϕ)-coordinate map of the sphere provides a single coordinate system for sensory data and control guidance angles. Thus, sensory data may be directly used for the control guidance system.

In this paper, the above result is further extended in two directions: (1) active sensing, and (2) coordination of active sensing and motor control. Active sensing means that specific advantages are obtained through sensor movements in the process of problem solving of complex tasks. In particular, active vision shows that knowledge of camera motion may stabilize a wide range of computations in early vision. For example, eye movements are commonly used by human beings in understanding our surroundings. Although passive sensing and motor control have been studied extensively, the coordination of active sensing and motor control is emerging to be a key question in complex robotic systems.

This research has been supported partially by the National Science Foundation.

Active sensing has two stages: (1) the control and coordination of multisensors, and (2) the adaptive computation in low level image or sensory data processing. In both stages, the adaptive approach of neural networks is very useful. The distributed and adaptive control of multisensors in the overall system architecture is performed by a neural network controller. The adaptive computation of images gathered by active sensors may also be represented by the neural network computational models [3],[4],[5]. The last but not the least is the adaptive coordination of active sensing and motor control. All three components lead to a hierarchical structure of neural networks. The adaptive and learning capabilities of neural networks play a key role in this active sensing and motor control paradigm.

2. System architecture

The overall system consists of multisensor systems which are located on either mobile robots or fixed sites. Multisensor families may be homogeneous or heterogeneous. Multisensor fusion may be classified into two categories - static and dynamic multisensor fusion [1]. Static multisensor fusion deals with the situation that both the sensors/robots and the environment are static. Although this case is not the main interest of this paper, it is basic to the understanding of the theory and implementation. Dynamic multisensor fusion is concerned with the scenario of a distributed system of mobile robots equipped with sensors and surveillance stations, exploring and understanding the environment, carrying out tasks, and communicating sensory data among themselves.

The hierarchical control scheme, depicted in the following, has adaptive and distributed control and active image processing capabilities. At the top level, a neural network controller provides the overall control of the complex sensor/robot system which is decomposed into several control subsystems. At the second level, these control subsystems have the responsibilities of guidance control and active sensing. The coupling of subsystems is performed at the top level. At the bottom level, tasks are actual sensing and image data processing. Thus, active sensing involves with the lower two levels; and the coordination of active sensing and motor control involves with all three levels.

An experimentation is planned for a mobile robot platform with multiple CCD cameras and ultrasonic sensors. This testbed is expandable to a distributed system of sensor/robots.

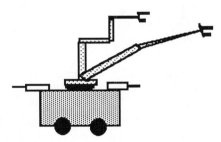

A Mobile Robotic System

3. Viewer-centered spherical model

A dynamic viewer-centered spherical model is located at the origin O of a moving coordinate system R^3. This represents a sensor/robot -centered coordinate system. In the next section, we shall discuss an object-centered coordinate system. As a sensor at the origin O rotates, the viewing sphere gives such a model. The usual image planes at different viewing angles are tangent planes to the sphere at different viewing axis. This is one kind of sensory data fusion on each viewing sphere. Moving targets are captured by such spheres.

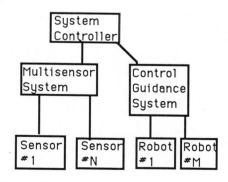

Overall System Architecture

The coordinate system $\{(\theta,\phi) : 0\leq\theta\leq\pi, 0\leq\phi\leq 2\pi\}$ of elevation and azimuth angles is a parametrization of the viewing sphere which we may call the retinal sphere of a family of sensors located at the origin O. As the sensor family moves, the retinal sphere undergoes a rigid motion M in the euclidean 3-space. Within a certain degree of accuracy (e.g. mobile robot slippery and miscalibration [6]), two retinal spheres are related by such a rigid motion M.

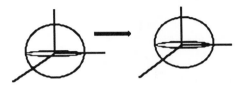

A Dynamic Spherical Coordinate System

The spherical model integrates image frames around a viewer. This is useful to understanding the environment around the viewer. For several viewers (or sensor families) not densely located, different sphercial images give quite different information about the global environment. In order to obtain more concrete and improved results, focus of attention to a particular object or a group of objects is necessary. The object-centered spherical model will be discussed in the next section for multiframes integration.

Another aspect of the dynamic spherical model is the capability of scaling (zoom-in and zoom-out) in different resolutions by moving the sphere along a viewing axis. In this case, we are really working with a focus-of-attention sector of the sphere. The remaining part of the sphere is not essential. Then, the part of the sphere cut out of the sector may be approximated by an usual image frame. However, two sectors are still conveniently related by a rotation on the sphere.

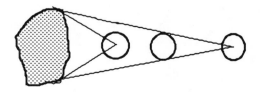

Scaling due to the motion of a sensor

4. Object-centered spherical model

In the last section, we have suggested that multiframes integration should be devoted to a particular object or a group of objects which is the focus of attention. In order to sense the occluded portion of an object, several sensor families

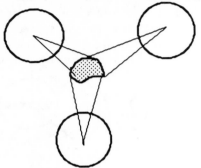

Focus of Attention

located at quite general positions are used to obtain multiframe views. The piecing together of these information is performed on an object-centered spherical model. The relationships among the viewer-centered and object-centered spheres are described in the following diagram:

In the literature, multiframe views are orthographically projected. Different frames are merged into a 3-D representation of the object. The object-centered spherical model offers a more efficient way to integrate different frames on the sphere. Moreover, it models the perspective projection rather than the orthographic projection. How do we integrate different frames on the object-centered sphere? First, the centroid or any interior point of an object is chosen as the center of the sphere. Since we do not have exact information of the object, the center is estimated within a certain degree of error. For example, a stereo vision algorthm will be sufficient to estimate the location of the origin as well as the distance to each viewer. Next, different concentric spheres with the estimated origin will intersect different viewers so that the relative position of any two frames in angles (Θ, Φ) of the object-centered sphere is determined. Further rotations are needed to align these frames in the spherical (ρ, Θ, Φ) grid. Information of each visible surface patch of the object is encoded in the object-centered octree structure.

This organization of spatial information of the object tells us quite accurately how much do we know about the object. If the sphere is not covered appropriately by enough sensors, some movements of the sensors will provide additional frames to fill in the gaps.

5. Active sensing

Recently, active vision and neural processing have emerged to remedy the deficiency of passive vision and image processing [7], [8],[9],[10],[11]. In [8], D. H. Ballard argued: " At first, the ability to make different kinds of programmed eye movements might seem to complicate a computational model, but in fact, the converse is true. The ability to make eye movements actually simplifies the information-gathering process of the eye."

In the following, we summarizes some of the points about active sensing:

(1) Active sensing provides additional constraints in the low-level computation of features in sensory data. It is known that optical flow and relative depth are calculated in a simpler manner than passive vision. Visual image understanding results are improved by active sensing [7], [8], [9], [10], [11]. The same situation holds for other types of sensing- thermal, radar, laser, acoustic sensors. For example, active range sensing is used to probe an object at a finer resolution for object recognition. Also, probing the object surface gives a complete description of the object.

(2) The correspondence (registration) problem may be solved by active sensing. Active sensing allows more efficient pixel-based correspondence by stochastic relaxation methods. Furthermore, correspondence may be accomplished by feature-based correspondence. Different features may be selected adaptively for optimal matching.

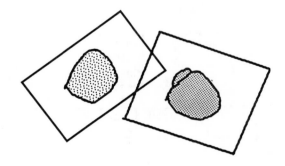

Correspondence
(Registration)

(3) Since multisensor fusion is a generalized problem of the correspondence problem, active sensing plays a similar role in the fusion problem. Namely, it allows more efficient pixel-based fusion by stochastic relaxation methods and permits feature-based fusion through actively selecting features for optimal matching.

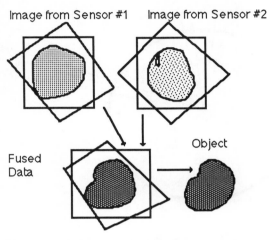

Multisensor Fusion

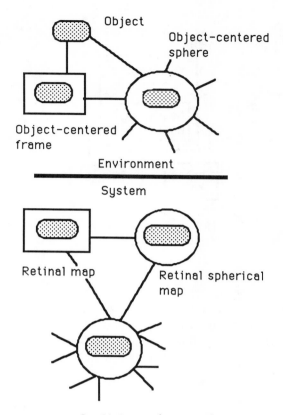

Spatial map (memory)

(4) Aloimonos et al [7], Ballard [8], Hinton [9], Kosslyn [10] and Marr [11] have investigated spatial relationships through active vision. The sequentiality of eye movements is used to encode spatial relations economically. Visual sequential search solves also the indexing problem. These work will bound to influence future research in spatial reasoning significantly.

These ideas may be enhanced by the neural network approach. The learning capability enables the system to do tasks that can not be done by traditional techniques.

6. Adaptive control by neural networks

In the above, we have described some ideas of active sensing and active low-level computer vision (or image processing). The main question are what are the control strategies of active sensing and how to implement them, including the control strategies of sensor movements and visual /sensing search. In the hierarchical structure proposed in section 2, the middle level is addressing this question.

Sensor movements are controlled by the middle level neural network controllers. Inputs to a neural network controller are different sensory images. There is a feedback loop from the bottom

level sensor systems to a given controller. Several measures may be defined to evaluate the preliminary multisensor fusion results. One obvious measure is the difference of pixel matchings. These measures are represented by node values of certain hidden units (the evaluation matrix) of the network. In the network, the sensory memory is a spatial map of the environment where different image frames are

Image Inputs from Sensors

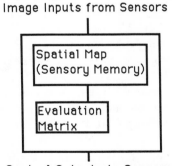

Control Outputs to Sensors

Active Sensing: Sensor-Control

fused. It is generally recognized that fusion should be done at the 3-D spatial level rather than at the image level. So, the spatial map should contain spatial 3-D information.

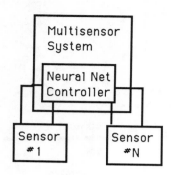

Multisensor System
Feedback Loops

10. S. Kosslyn, Seeing and imagining in the cerebral hemispheres: A computational approach, Psych. Review, 94, 1987, 148-175.
11. D. Marr, Vision, W. H. Freeman, 1982.

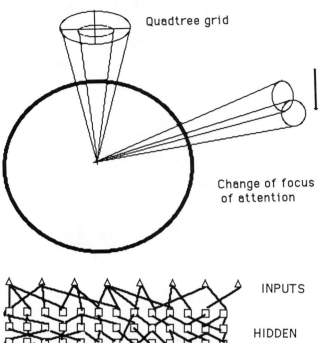

Quadtree grid

Change of focus
of attention

Sensor movements for fusion are determined by the evaluation matrix of matching and correlation measures. For spatiasl relation and visual/sensing search, the evaluation matrix is given by measures of states in search spaces.

References

1. S. Chen, Multisensor fusion and navigation of mobile robots, International Journal of Intelligent Systems, Vol. 2, 1987, 227-252.
2. S. Chen, A geometric approach to multisensor fusion, Proc. AAAI Spatial Reasoning and Multisensor Fusion Workshop, October 5-7, 1987, 201-210.
3. S. Chen, Neural networks and compuer vision, Proc. IEEE First Annual International Conference on Neural Networks, San Diego, 1987.
4. D. Rumelhart, J. McClelland and the PDP Research Group, Parallel Distributed Processing, Vol. I and Vol. II, The MIT Press, 1987.
5. G. Hinton and J. Anderson (Ed.), Parallel Models of Associative Memory, Lawrence Erlbaum Associates, 1981.
6. B. Kuipers and Y. Byun, A qualitative approach to robot exploration and map-learning, Proc. AAAI Spatial Reasoning and Multisensor Fusion Wirkshop, October 5-7, 1987, 390-404.
7. J. Aloimonos, I. Weiss and A. Bandopadhay, Active vision, Proc. International Conference on Computer Vision, London, 1987, 35-54.
8. D. Ballard, Eye movements and visual cognition, Proc. AAAI Spatial Reasoning and Multisensor Fusion Workshop, October 5-7, 1987, 188-200.
9. G. Hinton, Shape representation in parallel systems, Proc. 7th IJCAI, Vancouver, B.C., August 1981, 1088-1096.

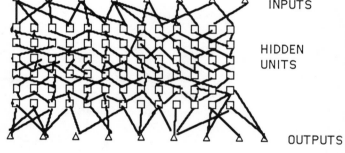

INPUTS

HIDDEN
UNITS

OUTPUTS

A Neural Network

Neural network approach to sensory fusion

John C. Pearson[1], Jack J. Gelfand[1], W.E. Sullivan[2]
Richard M. Peterson[1] and Clay D. Spence[1]

1. David Sarnoff Research Center, Subsidiary of SRI International,
Princeton, New Jersey 08543-5300
2. Department of Biology, Princeton University,
Princeton, New Jersey 08544

ABSTRACT

We present a neural network model for sensory fusion based on the design of the visual/acoustic target localization system of the barn owl. This system adaptively fuses its separate visual and acoustic representations of object position into a single joint representation used for head orientation. The building block in this system, as in much of the brain, is the neuronal map. Neuronal maps are large arrays of locally interconnected neurons that represent information in a map-like form, that is, parameter values are systematically encoded by the position of neural activation in the array. The computational load is distributed to a hierarchy of maps, and the computation is performed in stages by transforming the representation from map to map via the geometry of the projections between the maps and the local interactions within the maps. For example, azimuthal position is computed from the frequency and binaural phase information encoded in the signals of the acoustic sensors, while elevation is computed in a separate stream using binaural intensity information. These separate streams are merged in their joint projection onto the external nucleus of the inferior colliculus, a two dimensional array of cells which contains a map of acoustic space. This acoustic map, and the visual map of the retina, jointly project onto the optic tectum, creating a fused visual/acoustic representation of position in space that is used for object localization. In this paper we describe our mathematical model of the stage of visual/acoustic fusion in the optic tectum. The model assumes that the acoustic projection from the external nucleus onto the tectum is roughly topographic and one-to-many, while the visual projection from the retina onto the tectum is topographic and one-to-one. A simple process of self-organization alters the strengths of the acoustic connections, effectively forming a focused beam of strong acoustic connections whose inputs are coincident with the visual inputs. Computer simulations demonstrate how this mechanism can account for the existing experimental data on adaptive fusion and makes sharp predictions for experimental test.

1. INTRODUCTION

Neural network research is largely oriented towards solving "high-level" problems such as pattern recognition, categorization, and associative memory. Recent developments in this field have produced powerful new architectures and algorithms for solving such problems (see review by Cowan and Sharp[1]). Due to our relative ignorance of how such problems are solved by animals, these "high-level" neural networks are only loosely based upon known principles of brain organization and function, and do not directly correspond to any known brain structures. However, there are other equally important problems for which the corresponding brain structures are well characterized. Object localization and identification is one such problem, and sensor fusion plays an important role in the brain's solution to this problem. In this paper, a neural network approach to sensor fusion is presented that is based upon the map-like brain structures that solve the acoustic object localization problem in the Barn Owl.

Neuronal maps are key building blocks of nervous system function, ranging from perceptual classification to motor control. These structures consist of locally interconnected arrays of neurons whose response properties vary systematically with position in the array, thus forming a map-like representation of information. Computation is achieved through transforming the representation from one map to the next. The fidelity of these transformations is maintained through dynamic processes of self-organization, endowing them with self-optimizing and fault tolerant properties. These structures are linked together in modular, hierarchical processing systems, that employ some of the same problem solving approaches used in technical applications, such as sensor fusion.

In this paper we first briefly describe the stages in the chain of neuronal processing that generate a map of space from acoustic timing cues, and adaptively fuses it with the map of space derived from the retina. We then describe our proposed neuronal mechanism for the stage of visual/acoustic fusion and present results of computer simulations. This mechanism exploits the coincident signals produced by an object in the visual and acoustic representations, using adaptive, non-linear, neuron-like processing elements.

2. BARN OWL VISUAL/ACOUSTIC FUSION SYSTEM

Behavioral and physiological studies have revealed that owls use interaural intensity cues to specify the elevation of sounds, and interaural timing cues to localize the azimuthal direction[2,3]. The neuronal processing leading to azimuthal sound localization and visual fusion is accomplished by a series of four so called "computational maps", as shown in Figure 1, and reviewed by Knudsen[4]. The processing for elevation follows a similar design and is omitted from the present discussion for clarity and brevity.

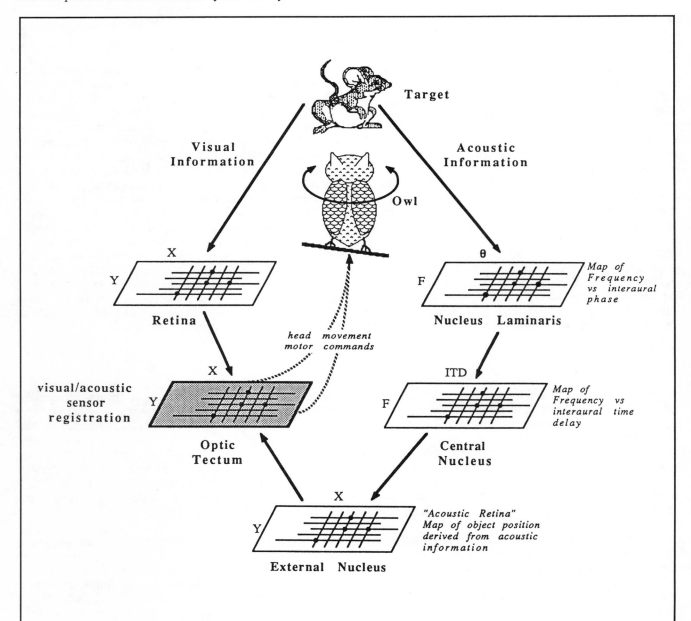

Figure 1. Illustration of the series of transformations in neuronal representation that produce adaptive visual/acoustic fusion in the object localization system of the Barn Owl. Only the azimuthal acoustic system is shown for clarity. A similar parallel series of transformations between neuronal arrays computes elevation.

Nucleus laminaris (N. lam.) generates a map of interaural phase delay vs frequency given phase-locked input signals from the cochlear nucleus. The central nucleus of the inferior colliculus (ICc) transforms the N. lam. map into a map of frequency vs interaural delay. The external nucleus of the inferior colliculus (ICx) transforms the ICc map into a map of space, forming an "acoustic retina". The acoustic space map (ICx) and the visual space map (retina) are fused in their joint projection onto the optic tectum. This fused map of object location is then used in orienting the head to center the object in the visual field for closer scrutiny.

The neuronal processes of visual/acoustic fusion are known to be adaptive during the growth period of the owl[5]. This is essential for the young owl, for during this time the distance between the ears increases severalfold, and this distance is a critical parameter in computing azimuthal position. Laboratory experiments have shown that perturbations to either the visual or auditory transducers (e.g.. goggles or ear plugs) initially cause registration errors between the auditory and visual space maps. As a consequence, head orientation driven by acoustic cues fails to center the object in the visual center of view. However, with time, fusion is reestablished, and proper localization behavior is restored in a continuous manner.

The fused sensory map of space in the optic tectum is also a motor map that orients the head to center objects in the visual field, as illustrated in Figure 2. The static topographic projection of the retina onto the tectum, and the fact that the eyes do not move relative to the head, establishes head centered retinotopic coordinates on the tectum. Because the acoustic map (ICx) is fused with the visual map, the same region of the tectum will be activated by either visual or acoustic signals from a particular location in space. A vector from this point to the point representing the center of the retina represents the magnitude and direction of the head movement necessary to bring this source to the center of the visual field. Possible neuronal mechanisms for this have been presented by Grossberg and Kuperstein[6] for the similar case of saccadic eye movements. Multiple objects presumably create multiple regions of activation on the tectum, and the system must select a single target for the head orientation response. Didday[7] and Arbib[8] have presented neuronal models of this function for the case of the frog.

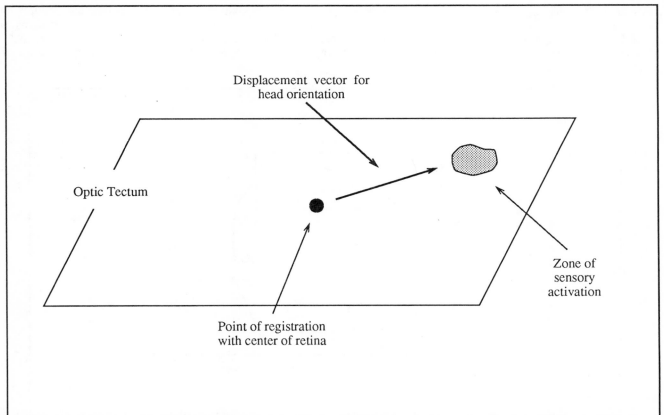

Figure 2. Representation of motor command for head orientation by position of activation on neuronal map in head centered retinotopic coordinates.

3. NEURONAL MECHANISM FOR SENSORY FUSION

Recent work by Pearson, Finkel and Edelman[9] demonstrated a solution to a problem related to the neuronal map fusion problem. They modeled the cortical map of touch sensation of the hand, which contains a topographic representation of both front and back hand surfaces. The representation of these two surfaces is not fused. Each cell in the map responds to stimulation of only one surface, and cells with the same preference are clustered into regions that are separated from regions with the opposite preference by sharp borders. Experiments have shown (see references within ref. 9) that the borders between these regions dynamically shift so that more highly stimulated regions of the hand have larger regions of representation in the map and greater resolution. In the model, each cell received equal numbers of connections from corresponding regions of both surfaces. A rule for changing the strengths of the connections strengthens inputs that are spatially and temporally correlated, and weakens those that aren't. Since the two surfaces are rarely stimulated at the same time, cells weaken their connections with one surface while strengthening their connections to a small, compact region of the other.

Our model for fusion in the tectum is a simplified version of this model of the map of the hand. The front and back surfaces of the hand correspond to the visual and acoustic space maps. Fusion is produced, instead of segregation, because stimulation of corresponding regions of the two input maps is correlated rather than uncorrelated. Figure 3 is a pictorial representation of the flow of signals from an object in the environment through the visual and acoustic space maps to a cell (marked with a filled dot) in the optic tectum. The light lines from the marked ICx cell to the tectum delimit its divergent region of projection, while the heavy line from the ICx cell to the tectum represents the functional projection created by strengthening that subregion of the total projection and weakening all others. This divergent projection is an assumption of the model. The dashed prism indicates schematically what would happen if there is a distortion added to the visual field. Immediately after the perturbation, the visual and acoustic maps in the tectum are out of register. The new point of activation in the retina (marked with an unfilled dot) immediately leads to the activation of a different cell in the tectum (marked with an unfilled dot) whereas the acoustic input fires the same cell in the tectum as before (filled dot). As a result, a single object will activate two cells in the tectum instead of one, and the input to a tectal cell will be half of what it was before the perturbation. However, with sufficient correlated stimulation of the visual and acoustic input maps, the connection strengths are altered so as to strengthen those acoustic connections that are coincidently activated with the visual input, and to weaken the original acoustic connections that are no longer activated at the same time as the visual connections. To test these ideas a simplified computer model was constructed.

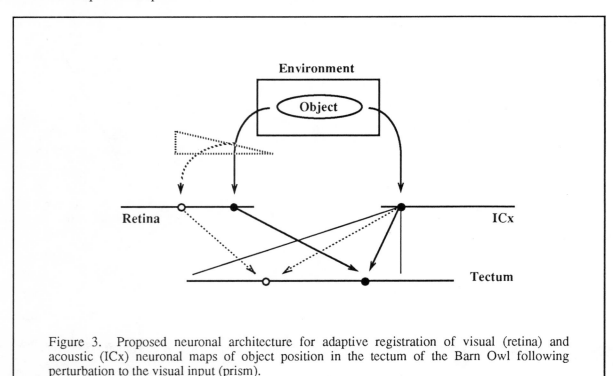

Figure 3. Proposed neuronal architecture for adaptive registration of visual (retina) and acoustic (ICx) neuronal maps of object position in the tectum of the Barn Owl following perturbation to the visual input (prism).

4. COMPUTER MODEL OF SENSORY FUSION

Figure 4 shows preliminary results of a simulation of adaptive fusion following the type of perturbation to the visual input described in Figure 3. These drawings are plots of the input connection strengths from the acoustic space map (ICx) onto the cell at the center of the fused map (tectum). The series of six drawings shows these connections as initially assigned (0), after 1000 time steps of unperturbed input, (1000), and at four successive times after the perturbation that moved the center of coincident input to the upper right as indicated by the arrow.

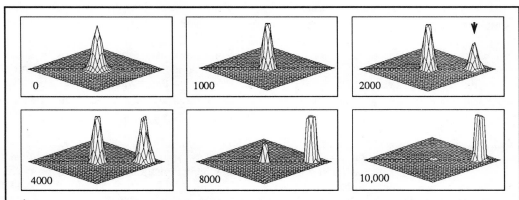

Figure 4. Simulation of reestablishment of visual/acoustic fusion following visual perturbation.. Arrow indicates site of acoustic input coincident with visual input after the perturbation.

In this simulation a visual/acoustic stimulus 3x3 grid points in size was applied in a random sequence over the entire input grid, coincidently activating topographically corresponding points in the space maps of the retina (R) and ICx (X). The tectal cell received a visual input from one fixed location (R_0) that was activated when the stimulus covered it. The stimulus generated input was summed to yield the cell's potential, v, shown in equation 1.

$$v = \sum_j c_j X_j + c_0 R_0 \qquad (1)$$

where:

c_j = variable acoustic connection strength, ($0 < c_j < 1$)

c_0 = fixed visual connection strength, ($c_0 = 10$)

X_j = acoustic input, ($X_j \varepsilon \{0,1\}$, 1=on, 0=off)

R_0 = visual input, ($R_0 \varepsilon \{0,1\}$, 1=on, 0=off)

The acoustic connection strengths were then modified according to a simplified version of the synaptic rule proposed in reference 9, shown in equation 2.

$$\Delta c_j = \kappa \cdot \sigma(v, \theta_1, \varepsilon_1) \cdot X_j - \delta \cdot \sigma(v, \theta_2, \varepsilon_2) \cdot |1 - X_j| \qquad (2)$$

where:

κ = growth constant, ($\kappa = .05$)

δ = decay constant, ($\delta = .05$)

σ = sigmoidal function, $\sigma(v, \theta, \varepsilon) = [\exp(\theta - v)/\varepsilon + 1]^{-1}$

θ = threshold parameter, ($\theta_1 = 9.5$, $\theta_2 = 16$)

ε = sharpness parameter, ($\varepsilon_1 = .01$, $\varepsilon_2 = .01$)

Parameters θ_1 and θ_2 set the thresholds for significant strengthening and weakening, respectively. Significant strengthening of an acoustic connection requires that its input be active ($X_j = 1$) and that enough other strong connections (whether visual or acoustic) be active so that $v > \theta_1$. Weakening an acoustic connection requires that its input be inactive ($X_j=0$) and that enough other strong connections be active so that $v>\theta_2$. Given the fixed value of c_0 (arbitrarily chosen) and the stimulus size (chosen based on simple considerations of scale), θ_1 was set so that weak acoustic connection strengths would grow very slowly unless they were activated coincidently with the visual input, and θ_2 was set so that only the coincident activation of both strong visual and acoustic inputs would produce weakening in the inactive acoustic inputs. Parameters κ and δ simply determine the time scale or "smoothness" of the simulation.

The reestablishment of fusion is robust to changes in the parameters, as long as the above guidelines are met. Larger stimuli simply enlarge the region of the strong acoustic connections. Larger values of ε soften the non-linearity in σ, making it easier to strengthen and weaken connections, but do not significantly affect the results. Regular stimulation, in which the stimulus moves over the input grid one point at a time, works as well as random stimulation. Changes in the threshold parameters affect the rate at which the original peak decays and the new peak grows, but not the final outcome, because the region of the new peak has the advantage of a fixed visual input. Of course, θ_1 must be less than the voltage due to the visual input, R_0, or the new peak cannot grow, and θ_2 must be less than the maximal potential.

The model makes several biological predictions that could be tested. The model assumes a topographic, divergent projection from the ICx to the tectum (see Fig. 3). This could be tested with various anatomical tracing methods. The width of the divergence sets the maximum range over which registration errors can be corrected. During the adaptation to the perturbation, the auditory responsiveness of cells in the tectum should change in a characteristic way. A new region of auditory responsiveness should appear along with the original region, and as the new region gains in strength the original region should weaken, eventually vanishing. At first, this appears to be in contrast with the behavioral result, which is that the localization error vector slowly decreases in magnitude. However, it is consistent with recent findings that the motor output is determined by the vector average of activity on the tectum[10].

5. CONCLUSIONS

Study of the visual/acoustic localization system in the Barn Owl has disclosed a potential neuronal mechanism for adaptive multi-sensor registration. Computer simulations of a neural network model of this system have successfully tested the proposed mechanism and produced predictions for experimental testing. Future work must determine the suitability of this method for technical applications.

6. ACKNOWLEDGEMENTS

This work was supported by internal funds of the David Sarnoff Research Center.

7. REFERENCES

1. J.D. Cowan and D.H. Sharp, "Neural nets", Daedulus, Winter (1988).
2. E.I. Knudsen and M.J. Konishi, "Mechanisms of sound localization by the barn owl (Tyto alba)", Journal of Comparative Physiology 133:13-21 (1979).
3. A. Moiseff and M.J. Konishi, "Neuronal and behavioral sensitivity to interaural time differences in the owl", Journal of Neuroscience 1:40-48 (1981).
4. E.I. Knudsen, S. du Lac and S.D. Esterly, "Computational maps in the brain", Annual Review of Neuroscience 10:41-65 (1987).
5. E.I. Knudsen, "Early auditory experience aligns the auditory map of space in the optic tectum of the Barn Owl", Science 222: 939-942 (1983).
6. S. Grossberg and M. Kuperstein, Neural dynamics of adaptive sensory-motor control, North-Holland (1986).
7. R.L. Didday, "A model of visuomotor mechanisms in the frog optic tectum", Mathematical Biosciences 30:169-180 (1976).
8. M.A. Arbib, "Visuomotor coordination: from neural nets to schema", in Adaptive Control of Ill-Defined Systems, Sulfridge et.al., eds., Plenum Press, New York, 207-225 (1984).
9. J.C. Pearson, L.H. Finkel and G.M. Edelman, "Plasticity in the organization of adult cerebral cortical maps: a computer simulation based on neuronal group selection", Journal of Neuroscience 7(12):4209-4223 (1987).
10. S. du Lac and E.I. Knudsen, "The optic tectum encodes saccade magnitude in a push-pull fashion in the barn owl", Society for Neuroscience Abstracts, 393 (1987).

Registration Error Analysis Between Dissimilar Sensors

Rae H. Lee [1,2] , W.B. Van Vleet [1]

[1] Rockwell International Corporation, Autonetics Sensors & Aircraft Systems Div.
Anaheim, CA 92803

[2] University of Southern California, Department of Electrical Engineering –System,
Signal & Image Processing Institue, Los Angeles, CA 90007

ABSTRACT

An approach is presented to estimate and minimize registration error between an optical or infrared imaging sensor and a nonimaging radar in a coboresighted multi–sensor object acquisition system. Previous efforts have been concentrated on analyzing the registration error between the sensed image and the reference image or between the same type of sensor signals. In this paper, models are developed for the centroid estimation error of an object from each sensor and for the registration error. The mean–square registration error is defined as a function of the sensor parameters and characteristics of the imaging sensor and the centroid uncertainty and pointing error of a monopulse radar. The optimal calibration range is estimated in terms of the sensor parameters and calibration object dimension.

1 INTRODUCTION

Registration between two sensors is the process of matching two sensor coordinate systems through a transformation to a common coordinate system and determining the relative offset (bias) between the two coordinate systems. The estimation of the offset provides informaion concerning the position of one sensor measurement relative to the position of the other sensor measurement. Since the measurements are never free of noise, a statistical method is needed to obtain an estimate of the true offset and the estimation error.

Most previous works related to registration error are concentrated on registration between the signals of the same type of sensors. Many of them are concerned with the problem of image–to–image registration by intensity correlation[1] , Fourier transform correlation[2] , or feature matching or symbolic registration techniques[3] . In reference 4, the registration error between two radars for a point object is analyzed by estimating the variance of the bias errors. Recently, multi–sensor systems (multiple and different types of sensor systems) have received attention in the automated object detection and guidance system environment because of their advantages, such as multi–spectral and multi–spatial perception, over single sensor systems. The registration error between dissimilar sensors has also received the attention lately as multi–sensor systems evolve from concept design to implementation.

In this paper, we consider the relative registration errors between an IR imaging sensor and a nonimaging monopulse radar when a hot corner reflector whose silhouette is convex is used as a calibration object. It is assumed that two sensors are colocated on a common gimbal so that there is no parallax error. (When two sensors are not colocated, the parallax error can be easily calculated from the distance between two sensors and the range and can be compensated.) Since the two signals from each sensor are quite different and there is no common feature that can be matched, the calculated centroids of objects are used as the registration control point in this effort. The centroid estimation models for each sensor are introduced and the registration errors are estimated based on the models and defined as a function of the relative angular separation between two neighboring pixels of an imaging sensor and the pointing error of a monopulse radar. The optimal calibration range is derived with known sensor parameters and physical dimensions of the calibration object.

2 CENTROID ESTIMATION ERROR MODEL IN AN IMAGING SENSOR

The centroid of an object in an image is usually calculated by averaging the pixel locations inside the silhouette (binary image) of the object. However, only the measurement noise of boundary pixels of the silhouette contribute error to the centroid estimate, while pixels inside the silhouette but not on the boundary do not contribute any error to the centroid estimate. In order to estimate the centroid estimation error, the centroid estimate for a silhouette image is viewed such that the centroid of the object silhouette is estimated by the weighted mean of the centroid estimates of each row and column of the binary image of the object.

The weights are the ratios of the number of pixels in a row (or column) divided by the total number of pixels inside the silhouette. The centroid estimate, (X_I, Y_I), in a silhouette image can be written as

$$X_I = \sum_{i=1}^{Nr} \frac{r(i)}{A} x_c(i) \quad \& \quad Y_I = \sum_{j=1}^{Nc} \frac{c(j)}{A} y_c(j) \tag{1}$$

where $x_c(i)$ is the i-th row centroid estimate, $y_c(j)$ is the j-th column centroid estimate, $r(i)$ is the number of pixels in the i-th row, $c(j)$ is the number of pixels in the j-th column, Nr is the number of rows, Nc is the number of columns, and A is the pixel area of the silhouette image.

2.1 Boundary measurement model

The object boundary measurement is modeled as the signal plus additive noise. The signal in this case is the true boundary position with respect to an imaging sensor's coordinate system plus the noise which is the spatial measurement error caused from the discrete scene sampling. The boundary measurements in a row of the silhouette image are modeled as follows:

$$z_l = x_l + n_l \quad \text{and} \quad z_r = x_r - n_r \tag{2}$$

where z_l and z_r are the left and right boundary pixels respectively, x_l and x_r are the left and right true boundary positions, and n_l and n_r are the random spatial noises which are independent of each other and also independent of x_l and x_r. In the image plane, the z_l and z_r are also the centers of the detector cells corresponding to the boundary pixels of the object.

The true boundary lies between a point inside the boundary detector cell, say Ca, and a point inside the nearest detector cell, say Cb, which is outside the silhouette. When all detectors are identical and the intensity slope of the local area of the boundary can be approximated to a linear slope, the spatial noises can be modeled as uniform random variables whose interval of distribution is identical to the distance between the center of Ca and the center of Cb, which corresponds to the relative angular separation between detector cells given the angle of the field of view (FOV) of the imaging sensor. These relationships are shown in Fig.1. Thus, the probability density function of the spatial noise is modeled as

$$f_{ni}(n) = \begin{cases} (Nx-1)/AFOV & \text{if } 0 < n < AFOV/(Nx-1) \\ 0 & \text{otherwise} \end{cases} \tag{3}$$

where AFOV is the azimuthal angle of the field of view of the imaging sensor, Nx is the number of detector cells which corresponds to the AFOV, and $i = r$ or l.

2.2 Centroid estimation error in imaging sensor

The centroid estimate of each row for the azimuth axis is written as

$$x_c(i) = [z_l(i) + z_r(i)]/2 \tag{4}$$

and the azimuth centroid estimate of the convex silhouette of the corner reflector can be written as follows by combining equations (2),(4), and (1):

$$X_I = \sum_{i=1}^{Nr} \frac{r(i)}{A} \frac{[x_l(i) + x_r(i)]}{2} + \sum_{i=1}^{Nr} \frac{r(i)t(i)}{A} = x_{CI} + n_X \tag{5}$$

$$t(i) = [n_l(i) - n_r(i)]/2$$

where x_{CI} is the true azimuth centroid measurement of the corner reflector with respect to the imaging sensor coordinate system and $t(i)$ is the spatial random noise of centroid estimation of each row. The probability density function is identical to the correlation of two identical uniform density functions and results in a triangle density function as follows

$$
f_t(t) = \begin{cases} 2[1 - 2t/AS]/AS & \text{if } 0 < t < AS/2 \\ 2[1 + 2t/AS]/AS & \text{if } -AS/2 < t < 0 \\ 0 & \text{otherwise} \end{cases} \tag{6}
$$

$$
AS = AFOV/(Nx-1).
$$

Since the noises are independent, the total azimuth centroid estimation noise is the linear convex combination of the independently identically distributed (i.i.d.) random noises, $\{t(i)\}$, in (5). The linear convex combination means

$$
\sum_{i=1}^{Nr} \frac{r(i)}{A} = 1 \tag{7}
$$

Thus, the variance of the total noise, n_{Ix}, is

$$
\sigma_{Ix}^2 = Var\{ \sum_{i=1}^{Nr} \frac{r(i)}{A} t(i) \} = \sum_{i=1}^{Nr} \frac{r(i)^2 * AS^2}{48 * A^2} \tag{8}
$$

Since $t(i)$ is a continuous random variable with finite variance, and each density of $t(i)$ is zero for $|t(i)| > AS$, where AS is finite, and the resulting variance is finite because of convex combination, the distribution of the total centroid estimation noise, n_x, converges to the Gaussian by the central limit theorem.[5] Therefore, the total IR noise can be approximated by a zero mean Gaussian random variable whose variance is σ_{Ix}^2.

3 CENTROID MEASUREMENT ERROR MODEL IN A NONIMAGING RADAR

Centroid measurement error in a non-imaging radar comes from many sources, such as thermal noise, clutter, multipath reflections, target glint, quantization error, servo and mechanical error, and atmospheric propagation. The thermal noise and target noise (stationary target glint) are the main error sources in registration calibration. Frequency agility can be exploited to reduce the target noise. The centroid of an object is estimated by the weighted mean of the centroid measurements of each radar return scatterer which is distributed in K frequency steps. Many weighting functions, such as sample mean, linear weighting, square-law weighting, Rayleigh weighting, exponential weighting, and selected largest amplitude, were introduced in reference 6. The sample mean is used in this paper because of the uniform response from the calibration corner reflector.

3.1 Centroid measurement error model in a nonimaging radar

The centroid measurement of each radar scatterer is modeled as the true centroid with respect to the radar plus two independent zero mean noises. The azimuth centroid measurement of the k-th scatterer is modeled as

$$
z_x(k) = x_{CR} + v_x(k) \quad ; \quad v_x(k) = v_o(k) + v_p(k) \tag{9}
$$

where x_{CR} is the true azimuth centroid with respect to the radar coordinate system, $v_p(k)$ is the pointing error of the monopulse radar due to thermal noise, and $v_o(k)$ is the error corresponding to the target noise, the centroid uncertainty of the object due to radar characteristics at k-th frequency. A target noise model can be complex in general. But the glint error of a stationary corner reflector can be modeled as a zero mean Gaussian random variable whose 3-sigma value is the angle corresponding to the physical limit of the object so that its standard deviation is

$$
\sigma_{vo} = OBL/6R \quad \text{for all K measurements} \tag{10}
$$

where OBL is the azimuthal length of the object and R is the range to the corner reflector.

The angle error due to receiver thermal noise of a monopulse radar is a function of many radar parameters. The variance of the angle error in a basic form[7] used in this paper is as follows

$$\sigma_{vp}^2 = \frac{BW^2}{2Km^2 \, SNR} \quad \text{for each pulse} \tag{11}$$

where BW is the antenna 3 dB beamwidth, and Km is the normalized monopulse error slope. SNR is the signal to noise ratio per pulse and is given by

$$SNR = \frac{P*G^2*l^2*\sigma_T*Wp}{(4\pi)^3*NT*R^4} \tag{12}$$

where P is the transmitted power, G is the antenna gain, l is the wave length, σ_T is target radar cross section, Wp is the transmitted pulse width, and NT is the total receiver noise.[8]

3.2 Centroid estimation error in nonimaging radar

For the radar the centroid estimate is the sample mean of the K scatterer centroid measurements and is written as follows:

$$X_R = \frac{1}{K} \sum_{k=1}^{K} [x_{CR} + v_X(k)] = x_{CR} + v_X \tag{13}$$

$$v_X = \frac{1}{K} \sum_{k=1}^{K} [v_o(k) + v_p(k)]$$

Since the centroid measurement for each radar return scatterer can be assumed independent of each other if the total bandwidth is sufficient, (if the frequency change is enough compared to the object dimension), the variance of the noise v_X can be written as

$$\sigma_{RX}^2 = [\sigma_{vo}^2 + \sigma_{vp}^2]/K \tag{14}$$

Since the noise term in each measurement is an i.i.d. random variable, the sample mean converges to a zero mean Gaussian random variable for large K.

4 REGISTRATION ERROR ANALYSIS

4.1 Registration bias and error

The estimate of the relative registration bias for the azimuth axis between two sensors is the difference between two centroid estimates:

$$OS_X = X_I - X_R = (x_{CI} - x_{CR}) + (n_X - v_X) \tag{15}$$

where the first term is the true azimuthal registration bias between the two sensor coordinate systems and the second term is the random registration error for the azimuth axis.

Since the measurements by two dissimilar sensors are independent, the mean-square registration error for azimuth axis is

$$MSE_X^2 = \sum_{i=1}^{Nr} \frac{r(i)^2*AFOV^2}{48*A^2*(Nx-1)^2} + \frac{OBL^2}{36*R^2*K} + \frac{BW^2}{2*Km^2*SNR*K} \tag{16}$$

By applying the same argument to the elevation axis, we can obtain the mean-square registration error for the elevation axis as follows

$$MSE_y^2 = \sum_{i=1}^{Nc} \frac{c(i)^2*EFOV^2}{48*A^2*(Ny-1)^2} + \frac{OBH^2}{36*R^2*K} + \frac{BW^2}{2*Km^2*SNR*K} \tag{17}$$

where EFOV is the elevation angle of the FOV, Ny is the number of detector cells which corresponds to the EFOV, and OBH is the height of the object. Notice that all the components of the mean-square registration error are known or measurable parameters from the physical dimensions, the characteristics of both sensors, and the calibration object.

4.2 Optimal calibration range estimation

The optimal range that minimizes the mean-square registration error is obtained by finding the range which satisfies the following equations:

$$\frac{d\,MSE_x^2}{dR} = 0 \quad \text{and} \quad \frac{d\,MSE_y^2}{dR} = 0 \tag{18}$$

and the optimal range that minimizes the azimuth registration error can be estimated as

$$Rop = \left[\frac{(C1^2 + 32*C2*C3)^{1/2} - C1}{8*C3}\right]^{1/3} \tag{19}$$

$$C1 = AFOV^2 * EFOV/[48*(Nx-1)^2 *(Ny-1)*OBH]$$

$$C2 = OBL^2/(36*K)$$

$$C3 = (4\pi)^3 * NT*BW^2 /[2*Km^2 *K*P*G^2 *l^2 *\sigma T*Wp]$$

where the shape of the corner reflector is approximated as rectangular.

5 DISCUSSION AND CONCLUSION

The intent of this paper is to analyze the registration error between dissimilar sensors, especially an imaging sensor and a nonimaging radar, in a multi-sensor system. In this paper we introduced the centroid estimation error model for the imaging sensor and an approach that estimates and minimizes the relative registration error between the imaging sensor and the nonimaging monopulse radar. With this result, given the sensors and the calibration object, one can estimate the relative registration error in terms of the sensor parameters and the dimensions of the calibration object. The registration error estimate is an important factor in determining the size of spatial correlation gates for a multi-sensor detection system.

6 ACKNOWLEDGEMENT

The authors wish to thank Dr. R. Zamow and Mr. S. Burrows in Rockwell International Co. for their review and comments on this paper.

7 REFERENCES

1. W.K. Pratt, "Correlation techniques of image registration," IEEE Trans. AES, May 1974, pp353-358.
2. P.E. Anuta, "Spatial registration of multispectral and multitemporal digital imagery using fast Fourier transform techniques," IEEE Trans. Geoscience Electronics, Oct. 1970, pp353-368.
3. K. Price and R. Reddy, "Matching segments of images," IEEE Trans. PAMI, Jan. 1979, pp110-116.
4. W. Fischer, C. Muehe, and A. Cameron, "Registration errors in a netted air surveillance system," MIT Lincoln Lab. TN 1980-40, Sept. 1980.
5. A Papoulis, Probability, Random Variables, and Stochastic Process, McGraw-Hill (1965).
6. J. Loomis and E. Graf, "Frequencey-agility processing to reduce radar glint processing error," IEEE Trans. AES, Nov. 1974, pp811-820.
7. E. Brookner, Radar Technology, Artech House (1977).
8. S. Hovanessian, Radar System Design and Analysis, Artech House (1984).
9. M. Skolnik, Introduction to Radar Systems, McGraw-Hill (1980).

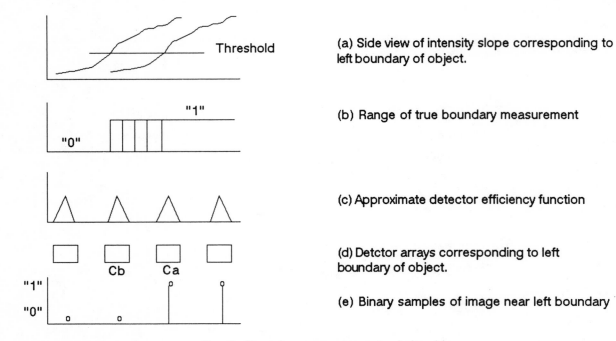

(a) Side view of intensity slope corresponding to left boundary of object.

(b) Range of true boundary measurement

(c) Approximate detector efficiency function

(d) Detctor arrays corresponding to left boundary of object.

(e) Binary samples of image near left boundary

Fig. 1 Boundary measurement relationship

Sensor/response coordination in a tactical self-protection system

Alan N. Steinberg

The Analytic Sciences Corporation, 8301 Greensboro Drive, McLean, Virginia 22102

ABSTRACT

This paper describes a model for integrating information acquisition functions into a response planner within a tactical self-defense system. This model may be used in defining requirements in such applications for sensor systems and for associated processing and control functions.

The goal of information acquisition in a self-defense system is generally not that of achieving the best possible estimate of the threat environment; but rather to provide resolution of that environment sufficient to support response decisions. We model the information acquisition problem as that of achieving a partition among possible world states such that the final partition maps into the system's repertoire of possible responses.

1. INTRODUCTION

1.1. Threat assessment/response system concept

The present application is that of developing an automatic planning function for coordinating the self-defense of a combat aircraft. A concept for a Threat Assessment/Response system architecture has been reported in Steinberg[11,12,13]. This concept has been developed as an independent research effort, in response to a generic need to develop technology for integrating system assessment, planning and control functions in tactical aircraft.

The objective of the Threat Assessment/Response system is that of maximizing the probablity of the aircraft surviving the mission while achieving mission objectives (defined below as "Mission Attainment Survivability"; abbreviated "MAS").

Key functions of the system are to (a) plan the assignment of sensors, counter-measures and evasive/avoidance maneuvers, given the uncertainties in threat assessments and in response effectiveness; (b) schedule actions within time constraints for measurement, countermeasure and processing; (c) coordinate the use of sensors to reduce uncertainties as required for response planning and cueing; and (d) resolve conflicting demands on assets so as to maximize global MAS.

Threat assessment functions (a) estimate threat identity, lethal potential and intent on the basis of available sensor and intelligence data and (b) predict the time to critical events in threat engagements (e.g. target acquisition, tracking, weapon launch, impact).

Response planning functions (a) develop candidate responses to reported world states; (b) estimate the effects of candidate actions on survival and on mission objectives; (c) identify conflicts for resource uses or detrimental side-effects of candidate actions; and (d) resolve conflicts to assemble composite Asset Assignment Plans, based on the estimated net impact on mission attainment survivability.

1.2. Information acquisition requirements

There are well-established procedures for defining requirements for defensive weapons and Electronic Warfare countermeasures and for evaluating their effectiveness.[10] The utility of such responses can be evaluated in terms of their impact on the platform's probability of survival and mission objectives.

However, the utility of sensor systems and sensor allocation planners in defensive systems can be related only indirectly to survivability and mission effectiveness factors.

In most cases, warning sensor requirements have been defined in a bottom-up manner; based on what is assumed to be the best achievable capability consistent with platform, cost and programmatic constraints. Thus requirements are levied for systems to be able to classify such-and-such a percent of all "high priority" emitters within x seconds and such-and-such a percent of all other emitters within y seconds.

In fact, such definitions of requirements are generally beside the point of the sensor's utility in mission. What counts is the ability to recognize a particular threat situation with sufficient accuracy to be able to select an appropriate response from the system's repertoire and to do so in enough time for the selected response to have a high likelihood of success.

Responses will be selected either to prevent transition to a more threatening state or to force a transition to a less threatening state. Accordingly, it is the role of sensors and associated situation assessment processing to (a) recognize particular world states which may lead to lethal states, (b) estimate the conditional likelihood of transitions to other particular states, given candidate action sequences and (c) estimate the time of such state transition.

The effectiveness of information acquisition is dependent not only on sensor performance, but also on the control system used for allocating sensors to specific tasks and on situation assessment processing (including fusion of data from multiple sensors). Therefore, for a given sensor/situation assessment system, information acquisition evaluation becomes a methodology for evaluating information acquisition planning and control strategies.

1.3.   The role of information acquisition in response planning

Our planning application differs from most which have been discussed[1,2,4,8,9] in that the world states are not generally fully resolvable by any information acquisition strategy available to the system. The effectiveness of a candidate response decision in influencing world state transitions depends to a certain extent on factors which will be outside the system's ability to predict. Therefore, the utility of a given response decision is generally uncertain.

Assuming perfect knowledge of current and future world states, a schedule of response actions could be defined which would be optimal, in terms of maximum probability of survival while achieving mission objectives, given the system's available repertoire of actions.

In practice, the system must generate response strategies based on the error-prone estimates provided by realistic sensors and associated processing and control. The performance of information-acquisition functions in this application must therefore be defined in terms of the degradation in MAS resulting from the decisions based on the estimates provided.

Because the system's certainty at any given time in predicting events will generally be greater with near-term events than with ones further in the future, tactical decision systems will tend to trade the possibility of incurring future penalties for a measure of near-term security.

In other words, uncertainty in threat assessment will tend to result in overreactions to the threat environment. In a tactical self-protection system, uncertainties in object classification, object state, location and in the time prediction of future events may result in such over-responses as:

    a.   inadvertently entering a threat system's lethal zone;

    b.   directing countermeasures against friendly systems;

    c.   using countermeasures against a threat while it is out of range, engaging someone else or being countered adequately by someone else;

    d.   using too much or too little jammer power or duty factor;

    e.   directing active E-O countermeasures against false alarms;

    f.   employing tracker or seeker deception techniques prior to lock on or after break-lock;

    g.   employing break-lock techniques too late for effectiveness;

    h.   using expendables or evasive maneuvers too early or too late for effectiveness.[3]

Such over-responses can jeopardize the aircraft's probability of survival and of achieving mission objectives in the following ways.

Excessive radiation or flight path change can increase the danger of tracking and multiple launches by given threats and can incur exposure to additional threats.

Directing sensors or countermeasures in ways which provide no benefit can reduce the system's ability to deal concurrently with threats: either by reducing the available assets or by interfering with other assets so employed.

Excessive expenditure of fuel, munitions or countermeasures stores, or of time in responding to perceived threats, can reduce the system's ability to deal with threats later in the mission or to achieve mission objectives.

2. MEASURES OF EFFECTIVENESS

Evaluating information acquisition requires determining the contribution which a given piece of information makes to the response decision process in a particular circumstance and on the conditional utility of each response decision available in that circumstance.

In our application, payoff and cost functions are defined in terms of impact on the survival of the aircraft and on the attainment of mission objectives. We employ a survivability model based on that developed by JTCG/AS, in which survivability factors are related to force-level measures of mission effectiveness.[5] In particular, we may identify payoff and cost factors in an offensive mission (e.g. air strike, offensive sweep, defense suppression) as measures of mission attainment survivability (MAS), by means of the following definitional hierarchy.

2.1. Encounter survivability

The likelihood of the aircraft surviving an encounter with a given threat system is given by

$$P_{s/e} = 1 - P_{tse} + P_{tse}(1 - P_{ssk})^m ; \qquad (1)$$

where

P_{ssk} = Probability of single shot kill;

P_{tse} = Probability of engagement by a threat system;
    = $P_{detect} \cdot P_{assign|detect} \cdot P_{track|assign}$;

m = Number of shots fired at aircraft by threat.

2.2. Sortie survivability

The likelihood of the aircraft surviving the mission is defined as a function of encounter survivability:

$$P_{s/s} = \pi_n \exp[-ZDE(1 - P_{s/e_n})] ; \qquad (2)$$

where

ZDE = Zone density effectiveness; i.e. the probability of the aircraft's encountering threat n (i.e. entering its lethal zone).

2.3. Mission attainment survivability

Mission attainment survivability relates the above survivability factors to the achievement of force-level mission objectives; i.e. the destruction of intended targets:

$$MAS = 1 - \sqrt{1 - [(1 - P_{s/s})/P_{k/s}](G_0/S_0)^2} ; \qquad (3)$$

where

$P_{k/s}$ = Probability of kill of threat target by aircraft per sortie;

G_0 = Threat force size to be destroyed (mission objective);

S_0 = Own aircraft force size.

3. MODEL OF INFORMATION ACQUISITION

3.1. State space and decision space

The goal of information acquisition in a tactical defense system is to provide resolution of that environment sufficient to support response decisions. As described below, we model the information acquisition problem as that of achieving a partition among possible world states such that the final partition corresponds to exactly one member of the system's repertoire of responses.

Moore and Whinston[8] define a model for developing an information acquisition strategy as an element of an adaptive planner.

The goal of the planner is modeled as that of choosing a sequence of information acquisition actions; i.e. a strategy α and a decision function $\delta : B \rightarrow D$, to maximize the expected net payoff $\Omega*$:

$$\Omega*(\alpha, \mathbf{B}, \delta) = \sum_{B \in \mathbf{B}} \sum_{x \in B} \Phi(x) w*[x, \delta(B), C(B)]; \tag{4}$$

where $\mathbf{B} = \{B_1, B_2, \ldots, B_q\}$ is an exhaustive partitioning of the set of possible world states X; $B_i \cap B_j = \emptyset$ for $i \neq j$;

$\Phi : X \rightarrow [0,1]$ is a probability density function, with probability distribution $\pi : P(X) \rightarrow [0,1]$ on the power set of X, $P(X)$, defined by $\pi(Y) = \sum_{x \in Y} \Phi(x)$ for $Y \subseteq X$;

D is the set of available response decisions among which the decision function δ must select;

$C : B \rightarrow \mathbf{R}$ is a cost function; $C(B) = \sum_{t=1}^{r} c[a(t,B)]$, where $c[a(t,B)]$ is the cost of the tth action in the sequence leading to partition B;

$\mathbf{R}$ is the set of possible utility results;

$w* : X \cdot D \cdot \mathbf{R} \rightarrow \mathbf{R}$ is a payoff function.

3.2. Information structures and feasible strategies

Each information acquisition action $a \in A$ (e.g. sensor assignment, parameter measurement, signal sorting/correlation, signal or object classification task) will yield an information set (or "signal set") Y_a. In the ideal case of noiseless information we may represent the information results of such actions by means of a function $n_a : X \rightarrow Y_a$. Per Shafer, the belief function corresponding to n_a; $h_a : X \cdot Y \rightarrow [0,1]$; is "discounted" at a rate proportional to the *a priori* reliability of the sensor information. Moore & Whinston argue that the general case with noisy sensor information is reducible to a noise-free model.

An information acquisition action $\mathbf{a}$ induces an information structure $\mathbf{M_a}$ on X:

$$\mathbf{M_a} = \{M | \exists y[M = n_a^{-1}(\{y\})]\}. \tag{5}$$

A sequence of information acquisition actions creates a sequence of partitions on X, each a refinement on the previous partition in the sequence (i.e. each being at last as fine-grained as its predecessor). The refinement of an information structure $B = \{B_1, B_2, \ldots, B_k\}$ by an action function $\alpha : B \rightarrow A$ is defined as

$$R(B, \alpha) = \{r | \exists M(M \in \mathbf{M_a} \ \& \ r = B_j \cap M)\}. \tag{6}$$

A <u>feasible</u> information gathering strategy σ for D is a sequence of r partition/action pairs which culminate in a partition B_{r+1} which will support the response decision $\delta : B_{r+1} \rightarrow D$:

$$\sigma = \langle (B_1, \alpha_1), (B_2, \alpha_2), \ldots, (B_r, \alpha_r), (B_{r+1}, \delta) \rangle; \tag{7}$$

satisfying the following conditions:

(a) $B_1 = \{X\}$;
(b) $\alpha_t : B_t \rightarrow A$, $t = 1, \ldots, r$;
(c) $B_{t+1} = R(B_t, \alpha_t)$, $t = 1, \ldots, r$;
(d) $\delta : B_{r+1} \rightarrow D$;

Moore and Whinston show that this formulation of feasible strategy meets Bellman's condition for obtaining an optimal (maximum net payoff) solution via dynamic programming. Unfortunately, there is no generally effective procedure to work backward to develop an optimal strategy from a feasible set $B \subseteq X$ using induction.

Therefore, the planner may be obliged to use heuristics which will generally be suboptimal in resolving world states in support of a response decision.

4. INFORMATION ACQUISITION FUNCTIONS IN TACTICAL DEFENSE SYSTEMS

4.1. Tactical engagement models

Response planning involves reasoning in terms of the conditional utility of actions. We may represent encounters with various types of threat systems by means of finite state transition models such as depicted in Figure 1.

Each state in a tactical engagement model has an associated set of observable features. The system uses such state observable features in interpreting sensor data and in planning the use of sensor and countermeasure assets.

Each state transition in the model is associated with a set of necessary and sufficient conditions. These transition conditions allow prediction of the time to critical future states and are used in planning self-defensive responses: responses will be selected either in order to impede a particular state transition or to force a transition to a lower threat state.

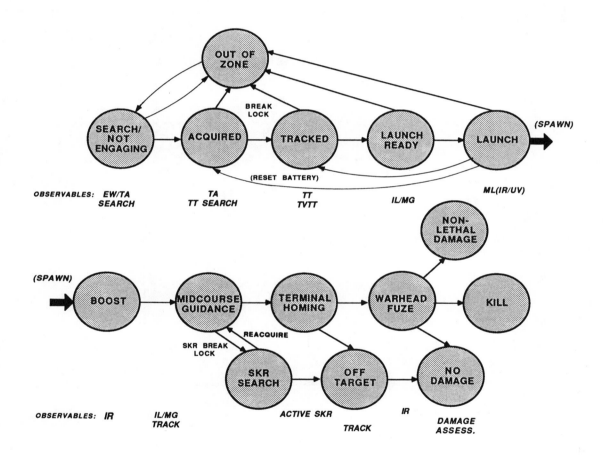

Figure 1. Tactical Engagement Model Concept

4.2. Survivability impact of information acquisition

A self-defense system performs the following information acquisition functions:

a. Threat detection;

b. Threat type determination (classification);

c. Threat state determination;

d. Object location/kinematics estimation;

e. Signal parameter measurement (for response technique definition and control and in support of other situation assessment functions);

f. Event likelihood prediction; i.e. predictions of state transitions in tactical engagement models; including estimation of P_s/e, modeled as the likelihood of avoiding a "kill" state;

g. Event time prediction;

h. Determination of own force disposition, state and predicted actions (for tactics coordination);

i. Determination of ambient visibility, atmospheric and terrain conditions (for planning response tactics and estimating sensor and response effectiveness).

We relate the contribution of these information acquisition functions to response decision factors in Table 1. These, in turn, affect mission attainment survivability factors as indicated in the lower half of the table.

Table 1. INFORMATION ACQUISITION EFFECTS ON SURVIVABILITY

INFORMATION ACQUISITION FUNCTION	RESPONSE DECISION FACTORS				
	Threat Avoidance	CM Assignment Targeting	Response Technique Selection	Technique Epoch Timing	Technique Parameter Control
a. Threat Detection	X	X	(X)	(X)	(X)
b. Threat Classification	X	X	X		X
c. Threat State		X	X	X	X
d. Location/Kinematics	X	X		X	2
e. Parameter Measurement					X
f. Event Prediction		X	X		
g. Event Time Prediction			X	X	X
h. Own Force Coordination		X	(X)		
i. Ambient Effect Est.	X		X		
SURVIVABILITY FACTORS					
P_{ssk} (single shot kill)		X	2	2	2
P_{detect}	X	(X)			
$P_{assign \mid detect}$					
$P_{track \mid assign}$		1	1	1	1
m (number of shots)	X	1		1	1
ZDE (prob. of encounter)	X				

X = primary effect   (X) = secondary effect   1 = pre-launch effect   2 = post-launch effect

5.   COSTS OF INFORMATION ACQUISITION

Since the indicated survivability costs of actions leading to a decision are linearly separable from the decision payoff, the net payoff of a decision strategy σ is the difference between the gross payoff and the cost of the actions:

$$\Omega*(\sigma) = \sum_{B \in B_{r+1}} \sum_{x \in B} \bar{\underline{q}}(x) w*[x, \delta(B)] - \sum_{B \in B_{r+1}} \pi(B) C(B); \tag{8}$$

where C(B) is the cost of all actions yielding partition B:

$$C(B) = \sum_{t=1}^{r} c[\alpha t (\beta t (B))];$$

βt(B) being the predecessor partition to B at step t; i.e. βt(B) is that B'∈B such that B∩B'≠∅.

Cost factors involved in information acquisition in an aircraft self-protection system will include:

a. Increased exposure to threats (i.e. increased ZDE) due to flight path changes in support of information acquisition; e.g. pop-up maneuvers for threat/target detection or for azimuth/elevation passive location; or delaying response decision until a sufficient baseline has been flown for triangulation location of a given accuracy;

b. Use of actively radiating information sources (e.g. radar, ladar, IFF interrogator, data link) incurring an increased likelihood of acquisition, tracking and lock-on by threats; i.e. increased Ptse;

c. Reduced countermeasures effectiveness due to reduced duty factor, power level or spectral coverage to accommodate sensor look-through; allowing increased Ptse and number of shots in respect to pre-launch countermeasures and in Pssk for post-launch countermeasures.

d. Reduced ability to assess the threat environment due to preemption of sensors by cueing or directed search; which may adversely affect the above three factors.

Figure 2 shows the causal interrelations among survivability factors, contributing to mission attainment survivability (MAS) and higher-level measures of mission success (MOMS). The effects of information acquisition cost factors are indicated at the left.

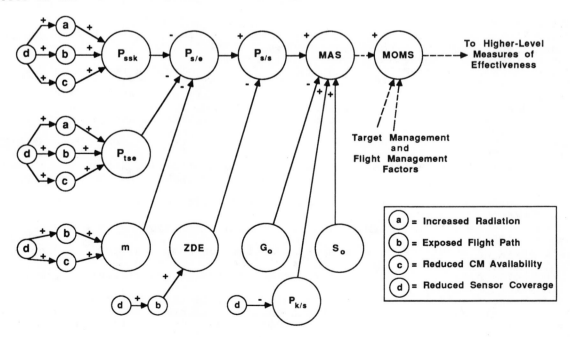

Figure 2. Survivability Factor Interrelationships

The payoff and cost of a response strategy may be defined in terms of its effect on mission attainment survivability:

$$\Omega*(\sigma) = (MAS|\sigma) - (MAS|\bar{\sigma});$$ (10)

which in turn is a function of the various factors which contribute to MAS:

$$(MAS|\sigma) = f(P_{ssk}, P_{tse}, m, ZDE, G_o, S_o, P_{k/s}|\sigma).$$ (11)

The payoffs and costs of particular information acquisition strategies may be estimated for a given system as a function of their expected impact on the survivability factors as shown in Figure 2 and Table 1.

The given model may be applied to a particular system by developing the applicable definition of specific functional relations corresponding to the top half of Table 1, between the system's repertoire of information acquisition actions and the specific decision factors involved in selection among the system's repertoire of responses.

The given model may be applied to a particular system by defining

a. the system's response decision function $\delta : B \rightarrow D$;

b. the system's payoff function $w* : X \cdot D \cdot MAS \rightarrow MAS$ in terms of specific response decisions $d \in D$ and threat states $x \in X$;

c. the likelihood of the planner selecting a feasible information acquisition strategy σ in situation x leading to a partition supporting the response decision δ per equation (7); and

d. the costs in terms of equation (11) of implementing strategy σ in situation x.

6. SUMMARY AND CONCLUSIONS

We have discussed the role which information acquisition plays in response planning in a tactical self-defense system. That role is one of partitioning the possible world state space in a way which allows mapping into the system's decision space. We have explored mechanisms for modeling this interrelationship and for evaluating the payoff and costs associated with information acquisition. Such a procedure, when completed by application to specific system performance and mission models, can be used in requirements definition, performance evaluation and in the design of adaptive Threat Assessment/Response Planners.

7. REFERENCES

1. Alterman, Richard. "An adaptive planner," _Proc. AAAI-86_ (1986): 65-69.
2. Dean, Thomas L. "Planning and temporal reasoning under uncertainty," _Proc. IEEE Workshop on Principles of Knowledge-Based Systems_, (1984): 131-137.
3. Doyle, Lawrence. personal communication (1986).
4. Doyle, R. J., Atkinson, D. J. and Doshi, R. S. "Generating perception requests and expectations to verify the execution of plans," _Proc. AAAI-86_ (1986): 81-88.
5. Drew, D. R., Trani, A. A. and Tran, T. K. "Aircraft survivability and lethality tradeoff Model (ASALT)," _Aircraft Survivability_, vol. 10, no. 2 (1986): 13-18.
6. Firby, R. James. "An investigation into reactive planning in complex domains," _Proc. AAAI-87_ (1987): 202-206.
7. Georgeff, Michael P. "The representation of events in multiagent domains," _Proc. AAAI-86_ (1986): 70-75.
8. Moore, James C. and Whinston, Andrew B. "A model of decision-making with sequential information acquisition, _Decision Support Systems_, vol. 2 (1986): 285-307; vol. 3 (1987): 47-72.
9. Morgenstern, Leora, "Knowledge preconditions for actions and plans," _Proc. IJCAI-87_ (1987): 867-874.
10. Schleher, D. Curtis. _Introduction to Electronic Warfare_. Artech House, Norwood, Massachusetts (1986).
11. Steinberg, Alan N. "Threat management system for combat aircraft," _Proc. Tri-Service Data Fusion Symposium, DFS-87_ (1987): 528-542.
12. Steinberg, Alan N. "A planner for threat assessment and response," _Proc. NAECON-88_ (1988): in print.
13. Steinberg, Alan N. "An expert system for multispectral threat assessment and response," _Applications of Artificial Intelligence V, SPIE Proc._ vol. 786 (1987): 52-62.
14. Steinberg, Alan N. "A closed-loop simulator for tactical aircraft systems," _Proc. NAECON-87_ (1987): 1034-1041

SENSOR FUSION

Volume 931

Session 2 (continued)

System Aspects

Chair
Alan N. Steinberg
The Analytic Sciences Corporation

IR & MMW sensor fusion
for precision guided munitions

J.A. Hoschette and C.R. Seashore
Honeywell Inc., Defense Systems Group

10400 Yellow Circle Drive, Minnetonka, Minnesota 55343

Abstract

Sensor fusion techniques are currently being investigated for potential application to a variety of smart weapon systems. Primary sensor candidates include active and passive millimeter wave as well as active and passive infrared configurations. Improved overall dual-mode sensor performance must be carefully traded off with hardware complexity, packaging challenges, and increased costs associated with the sensor fusion process. An additional factor of significance is the selection of sensor geometry, side-by-side, or common aperture.

Introduction

Autonomous, fire-and-forget weapon systems carrying single mode sensors have limited capability to select valid targets from within a matrix of natural and man-made objects, can be decoyed by reflective and emissive countermeasures and have little control over their terminal aimpoints. To overcome these inherent limitations, attention is being directed toward sensor fusion involving millimeter wave radar and passive infrared in side-by-side and common aperture configurations. Figure 1 illustrates several weapon system families that potentially can benefit from a fusion of spectrally separated sensors. The selection of millimeter wave and infrared sensors for such requirements requires considerable care since available packaging volume and aggressive cost goals are significant constraints to the hardware development process. Current attention has been largely directed toward a fusion between millimeter wave radar and passive infrared arrays with associated signal processing algorithms structured for optimal overall system performance.

The projected advantages from such a sensor fusion process include:

- Viewing a target scenario with multiple simultaneous sensor spectral wavelengths provides multiple discriminants and enhanced overall detection probability.

- Enhanced system operational effectiveness in an ever-expanding countermeasure environment.

- Rapidly advancing integrated circuit technology at millimeter wave and infrared wavelengths permit common aperture sensor realizations for small diameter weapon system requirements.

Sensors available for fusion into precision guided munitions include active and passive IR as well as active and passive MMW sensors. Each of these sensors have distinct advantages and disadvantages. The passive IR sensors offer hot spot detection and better aimpoint selection. However, they have poor smoke, fog, and rain penetration. The active IR sensor or laser rangers offer range to the target, target profiling, and precision standoff fuzing. Their chief disadvantage is also diminished smoke, fog, and rain penetration.

Passive MMW sensors or radiometers use natural radiation to seek out targets. They offer good weather penetration and passive detection but, however, encounter a lot of false targets. Active MMW sensors or radars offer good weather penetration, as well as range to the target and contrast data. However, their aimpoint resolution is somewhat less than that of IR sensors.

Sensor Fundamentals

Before discussing sensor fusion it is necessary to review sensor fundamentals. For this paper we will limit the discussion to passive IR and active MMW.[1,2] Shown in Figure 2 is a block diagram of an infrared sensor. The basic components include a collecting lens, optical passband filter, detector, detector cooler, and preamplifier. The analog-to-digital converter and signal processor are used to implement the target detection algorithms.

Infrared radiation from the target is collected by the lens and focused on to the detector. The optical passband filter eliminates all undesirable wavelengths. The detector converts the infrared energy to electrical signals. A thermo-electric cooler cools the detector to improve its signal-to-noise characteristics. The pre-amp provides signal gain and detector bias. The signal processor performs hot/cold spot detection and two-dimensional target detection in some systems.

Figure 3 shows block diagrams of radar sensors used in smart munitions and terminally guided munitions. The latter represents a significant hardware realization problem, particularly if the radar operating frequency is selected to be close to 100 GHz. In both cases, the RF lead contains an antenna and associated transceiver which make maximum use of integrated circuit subassemblies. The antenna can be either of a planar or a reflector design depending on the packaging volume available and the system performance desired.[3,4]

Sensor Fusion

Active/Passive MMW and IR sensors can be combined in several different configurations. For example, one may combine Active/Passive IR sensors or Active IR and Active MMW sensors. Each combination has unique advantages and disadvantages. The optimum combination will depend upon the weapon, environment to be used in, and target signature characteristics.

The predominate combination has been the fusion of an active MMW sensor with a passive IR sensor. This combination or dual mode operation offers very good performance for precision guided munitions. The IR sensor allows target detection based on hot spots, which are usually the more vulnerable areas of the target. The MMW allows target detection based on contrast, as well as range to the target. The MMW has better weather penetration which allows for longer range acquisition. Also, further aimpoint refinement can be made based on known MMW and IR centroids.

Figure 4 shows two methods for fuzing IR and MMW sensors. The first method is referred to as co-alignment. In this method, the IR and MMW sensors have a separate line-of-sight but are boresighted together. This method is easy to build, presently lower cost, and uses separate apertures.

The second method of fuzing the sensor uses a common aperture approach. Here a large primary reflector is used to focus both IR and MMW radiation down on to a secondary reflector. The secondary reflector is transparent to the MMW energy and highly reflective to the IR. The two signals are separated and sensed using separate detectors. The common aperture approach is harder to build, and further constraints the radome as it must pass both spectral bands.

Figure 5 summarizes the signal processing considerations for fuzing two sensors. A probability of detect (Pd) and false alarm (Pfa) can be computed for each sensor. However, the signal processor must make a decision based on two sensor inputs. If the sensor detects are simply ANDed together, this lowers the Pfa which is good but it also lowers the Pd. If the sensor detects are ORed together, this will increase the Pd but also increase the Pfa.

The real benefit of sensor fusion comes when combination logic is used in addition to simple ANDing or ORing. The combo logic may also look at other target signature characteristics such as size and signal strength. With combo logic based on known target signatures, it tends to hold Pd level or increase it and at the same time lower Pfa or hold it level when compared to single sensor operation.

In addition to choosing the right sensor combination to get optimum performance, one must consider hardware issues for each sensor. Consideration must be given for the MMW dome/rain cover, antenna configuration, and transceiver characteristics. In choosing the IR sensor one must consider the dome, array size, wavelength, and method of cooling. These hardware trades must be considered or the sensor cost may quickly become unreasonable.

System Applications

The weapon system applications which can benefit from a fusion of millimeter wave radar and passive infrared sensors have been generally described earlier in this paper. A partial set of functional requirements can be defined for such systems. These include the following:

- Autonomous fire-and-forget operation
- Capability for target acquisition, detection, and tracking
- Capability to resist active and passive countermeasures
- Hardware that is affordable and producible

- Operation in a gun-fired environment
- Hardware mechanization compatible with small diameter munitions
- System operability after up to 20-year storage life.

These are currently being addressed under several weapon development programs with technical results being very favorable in terms of future system implementations.

Conclusions

In summary, precision guided munitions represent a viable application base for sensor fusion including millimeter wave radar and passive infrared arrays. The critical sensor geometries include side-by-side and common aperture. A selection of the preferred geometry is based on available packaging volume and system cost goals. Several critical system issues currently being addressed during the development process include:

- Affordable and producible millimeter wave and infrared hardware
- Common aperture dome with multi-spectral response and survivability
- High environmental stress on hardware associated with gun-fired applications
- Severe packaging constraints associated with small diameter weapons.

References

1. Hoschette, J., Huffnagle, N., "Multifunctional Electro-optical Sensors for Precision Guided Munitions," 30th Annual SPIE, Orlando, FL, March 1987.

2. Huffnagle, N., Hoschette, J., Lubke, R., Ellingboe, J., "Active Laser Sensors for Submunitions," IRIS Specialty Group, John Hopkins University, November 1986.

3. Seashore, C.R., "Get Smart, Munitions - Move Up To Millimeter Waves," Defense Electronics, May 1984, pp. 118-133.

4. Seashore, C.R., "MIMIC for Millimeter Wave Integrated Circuit Radars," SPIE Volume 791, 22-21 May 1987, pp. 104-106.

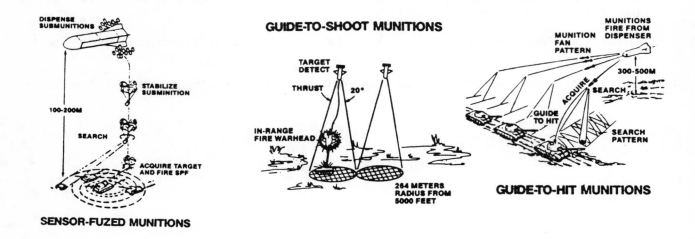

Figure 1. Weapon Systems Families

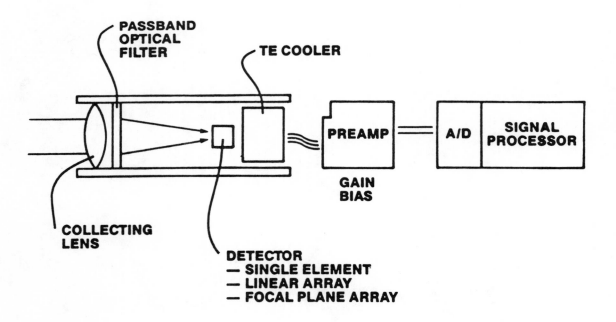

Figure 2. IR Sensor Block Diagram

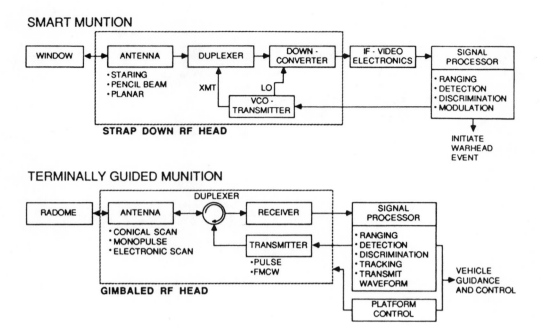

Figure 3. MMW Radar Sensor Block Diagram

COALIGNMENT

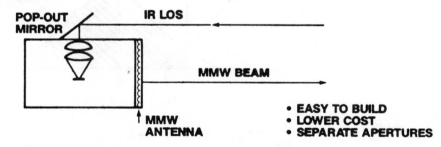

COMMON APERTURE

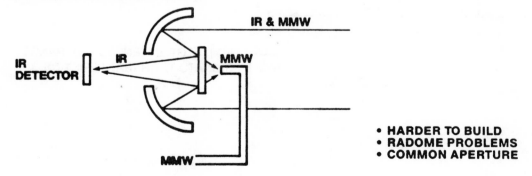

Figure 4. Sensor Fusion Methods

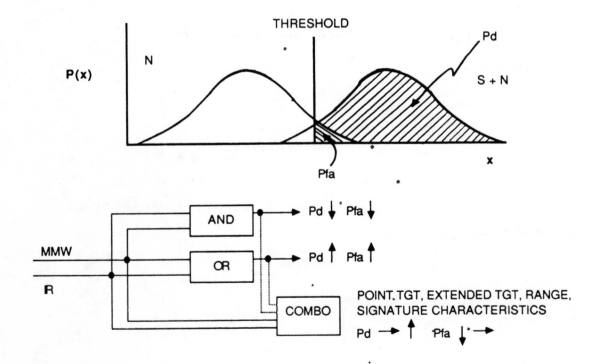

Figure 5. Signal Processing Considerations for Fuzing Sensors

SENSOR FUSION APPLIED TO SYSTEM PERFORMANCE UNDER SENSOR FAILURES

T. E. Bullock, S. Sangsuk-iam
Department of Electrical Engineering
University of Florida, Gainesville, Florida

R. Pietsch
LTV Missiles and Electronics Group
Dallas, Texas

E. J. Boudreau
U. S. Air Force Armament Laboratory
Eglin Air Force Base, Florida

ABSTRACT

System performance and reliability, given a sensor failure, can be improved by the use of multiple sensors of different types. A failure may be due to loss of information of any type. In order to use multiple sensors, it is necessary to perform local estimation, combine local estimations in a central processor, and to detect, isolate, and accommodate sensor failures. Methods for attacking these problems are generally known as sensor fusion. The general sensor fusion equations are set up showing how to construct local estimators, the central processing (fusing) algorithm, and "outer logic" for dealing with sensor failure. In addition application of the theory is demonstrated through simulation, with generic sensors employing non-linear models. The results show the detection, isolation and accommodation of sensor failures. Based upon this study, the concept of sensor fusion shows promise in making significant improvement in many systems employing multiple sensors.

I.   INTRODUCTION

System performance and reliability, given a sensor failure, can be improved by the use of multiple sensors of different types. Sensor failures may be due to actual failure of the sensor, or loss of sensor information due to any cause. Typical environments may degrade or blind a specific type of sensor and not disturb another sensor. In order to use multiple sensors, it is necessary to perform local estimation, combine local estimations in a central processor, and to detect, isolate, and accommodate sensor failures. Methods for attacking these problems are generally known as sensor fusion, although the term does not seem to have a precise definition in the literature.

In this paper we first set up the general sensor fusion equations, showing how to construct local estimators, the central processing (fusing) algorithm, and "outer logic" for dealing with sensor failure. Since applications are generally to non-linear systems, we consider local processors for non-linear models.

In order to demonstrate the application of the theory, simulation results with generic sensors, e.g., radar and IR, are given. Simulation results show the detection, isolation and accommodation of sensor failures. The advantage of sensor fusion are further demonstrated by a comparison of sensor performance with sensor failures when sensor fusion is used and when the local processors operate independently (no sensor fusion).

II.   THE CONCEPT OF SENSOR FUSION

Sensor Fusion in the initial study was defined as "combining the information from several different sensors to produce the single best estimate of the system state." It can be concluded that "Sensor Fusion" is Kalman filtering, modified for possible nonlinearities using the extended Kalman filtering and with added hypothesis testing for detection of an excessive noisy environment and for detection of an excessive noisy environment and sensor failures.

An initial premise is that missiles currently rely on only one sensor at a time to home in on the target. If this single sensor is lost due to excessive noise or failure, the missile cannot accomplish its mission. An obvious improvement is to use multiple sensors for redundancy in case of failure or sensors of different types which respond in different ways to a noisy and/or a deceptive environment. Although it may be clear how to combine a number of measurements using conventional filtering theory, the traditional way that missile sensors work for tracking is to only use one sensor type.

The Mathematical Problem

Using conventional filtering theory, the filter finds the "best" estimate of the n dimensional state x_k as a function of the measurements p dimensional $y_k, y_{k-1}, \ldots y_0$, and an assumed model

$$\hat{x}_k = E^* [x_k | y_k, y_{k-1}, \ldots y_0] \qquad (1)$$

The model used is a difference equation for the state

$$x_{k+1} = f(x_k, w_k) \qquad (2)$$

where w_k is the "system noise" sequence with known statistical properties. The measurement equation is

$$y_k = h(x_k, v_k) \qquad (3)$$

where v_k is the "measurement noise" sequence again with known statistical properties. Unfortunately, the results are strongest when linear. The notation which we will use for the linear case is

$$x_{k+1} = F_k x_k + w_k \qquad (4)$$

$$y_k = H_k x_k + v_k \qquad (5)$$

For convenience in describing the sensor fusion problem, we shall assume that there are two measurements, each a noisy measurement of only one of the states, each denoted by the superscript (1) or (2). Let superscripts be used to indicate rows of the x_k, y_k and v_k vectors. Then the measurements may be written as

$$y_k(1) = x_k(1) + v_k(1) \qquad (6)$$

$$y_k(2) = x_k(2) + v_k(2) \qquad (7)$$

Using these measurements, we find the "local" (1L),(2L) estimates

$$\hat{x}_k(1L) := E^*[\hat{x}_k(1) \mid y_k(1) , \hat{x}_{k|k-1}(1)] \qquad (8)$$

$$\hat{x}_k(2L) := E^*[x_k(2) \mid y_k(2) , \hat{x}_{k|k-1}(2)] \qquad (9)$$

where $\hat{x}_{k|k-1}(1)$ and $\hat{x}_{k|k-1}(2)$ are the a priori estimates of the first two components of the state vector using all of the measurements. Now we form the estimate

$$E^* [\hat{x}_k \mid \hat{x}_k(1L), \hat{x}_k(2L), \hat{x}_{k|k-1}] \qquad (10)$$

This procedure differs from the usual approach in that the measurements y_k are not directly used to estimate the full state vector. In the nomenclature of Sensor Fusion, the estimates $\hat{x}_k(1L)$, $\hat{x}_k(2L)$ are local estimates from local processors. The overall estimate given above is computed in the central processor.

Equations for Sensor Fusion

For a large number of definitions of the best estimate, the result is the conditional mean. The estimation problem may be solved by computing the conditional density,

$$p(x_k \mid y_k(1) , y_k(2)) = p(x_k(1), x_k(2) \mid y_k(1), y_k(2)) \cdot$$
$$p(x_k(r) \mid x_k(1), x_k(2)) \qquad (11)$$

where $x_k(r)$ is the state vector with the first two components omitted. This equation indicates the basic idea that the calculation of the state estimate can be split into two parts. One part is the local estimation, which corresponds to the first term. The second part is the central processing from the second term.

First we note that, in our example of the previous section, given $x_k(iL)$ and $\hat{x}_{k|k-1}(i)$ we can determine $y_k(i)$ since the estimation equation is a linear relationship between the variables which can

be inverted. Therefore the data set $\hat{x}_k(1L)$, $\hat{x}_k(2L)$ is equivalent to the data set $y_k(1)$, $y_k(2)$, i.e.,

$$E^*[x_k \mid y_k(1), y_k(2), \hat{x}_{k|k-1}]$$
$$= E^*[x_k \mid \hat{x}_k(1L), \hat{x}_k(2L), \hat{x}_{k|k-1}] \qquad (12)$$

The estimation equations in terms of the local estimates are simply derived by substitution of the measurements as a function of the local estimates into the equations for the estimates as a function of the measurements. In other words

$$y_k(i) = g(i)(\hat{x}_k(iL), \hat{x}_{k|k-1}) \qquad (13)$$

The "central processor" equation may be found by substitution of the $y_k(i)$'s as functions of the local estimates and a priori state estimates into the usual estimation equation

$$E^*(x_k \mid y_k(1), y_k(2), \hat{x}_{k|k-1})$$

The required error covariances may also be easily found using this approach. Using the linear measurement model (5), the local estimates are given by

$$\hat{x}_k(iL) = x_{k|k-1}(i) + K_k(iL) (y_k(i) - \hat{x}_{k|k-1}(i)) \qquad (14)$$

where $K_k(iL)$ is the optimal local estimation gain. The gain $K_k(iL)$ is given in terms of the local a priori covariance M_k by

$$K_k(iL) = M_k(iL) (M_k(iL) + R_k(i))^{-1} \qquad (15)$$

with the "local" a priori covariance $M_k(iL)$ expressed in terms of the overall a priori covariance M_k as

$$M_k(iL) = [M_k]_{ii} \text{ component} \qquad (16)$$

The optimal estimate of the state given the measurements is given by the familiar equations

$$\hat{x}_{k|k} = \hat{x}_{k|k-1} + K_k (y_k - H_k \hat{x}_{k|k-1}) \qquad (17)$$

with

$$K_k = M_k H_k' (H_k M_k H_k' + R_k)^{-1} \qquad (18)$$

The central processor using sensor fusion must find an estimate of x_k from $\hat{x}_{k|k-1}$ and local estimates $\hat{x}_k(iL)$. From (14)

$$y_k(i) = x_{k|k-1}(i) + \frac{1}{K_k(iL)} (\hat{x}_k(iL) - \hat{x}_{k|k-1}(i)) \qquad (19)$$

From (19), the desired sensor fusion central processor equations are found

$$x_{k|k} = x_{k|k-1} + \sum_{i=1}^{2} c_k(i) (\frac{1}{K_k(iL)}) (\hat{x}_k(iL) - \hat{x}_{k|k-1}(i)) \qquad (20)$$

where $c_k(i)$ is the ith column of K_k.

Although equation (20) has been derived for two measurements, similar results hold for more measurements. If some measurements are different measurements of the same variable, the term "sensor pooling" is used. In the case of different types of measurements, the term "sensor integration" is used.

Sensor Fusion for Correlated Measurements

The equivalent filter gain and measurement noise for the case of a vector measurement with $H=I$ are derived as follows. The covariance equation is

$$P = M - MH' (HMH' + R)^{-1} HM \qquad (21)$$

which reduces for $H=I$ to

$$P = M - K M \qquad (21\text{-}a)$$

with

$$K = M(M + R)^{-1} \qquad (22)$$

$$K = I - PM^{-1} \qquad (23)$$

and

$$R = K^{-1}M - M \qquad (24)$$

It is convenient to use sequential processing in place of batch processing for several sensors as it is then easy to bypass failed sensors and the check each sensor separately. To simplify the notation, consider two measurements

$$y_1 = H_1 x + v_1 \qquad (25)$$

$$y_2 = H_2 x + v_2 \qquad (26)$$

and let $\hat{x}_1$ be the estimate of x using y_1 and $\hat{x}_2$ be the estimate of x using y_1 and y_2. Furthermore, let P be the a priori covariance of x, P_1 be the covariance of the estimation error using $\hat{x}_1$ and P_2 be the covariance of the estimation error using $\hat{x}_2$. Then if $\text{Cov}(v_1, v_2) = 0$

$$\hat{x}_1 = \bar{x} + PH_1' (H_1PH_1' + R_1)^{-1}(y_1 - H_1\bar{x}) \qquad (27)$$

$$\hat{x}_2 = \hat{x}_1 + P_1H_2' (H_2P_1H_2' + R_2)^{-1}(Y_2 - H_2\hat{x}_1) \qquad (28)$$

In general the measurement noise terms are correlated so that the sequential equations are not correct. In this case batch processing can be used.

$$\hat{x} = \bar{x} + PH' (HPH' + R)^{-1}(y - Hx) \qquad (29)$$

$$P_2 = P - PH' (HPH' + R)^{-1}HP \qquad (30)$$

However, sequential processing may also be used with correlated measurement noise by using

$$\tilde{y}_{2/1} := y_2 - H_2 \hat{x}_1 \qquad (31)$$

as the second measurement in place of y_2 since $\tilde{y}_{2/1}$ is uncorrelated with y_1.

In this case the covariance of the measurement noise for the second measurement, which was R_2 in the case of $\text{Cov}(v_1, v_2) = 0$, becomes

$$R_2{}^n = \text{Cov} (\tilde{y}_{2/1}, \tilde{y}_{2/2})$$

We find

$$R_2{}^n = R_2 - H_2PH_1' (H_1PH_1' + R_1)^{-1}R_{12} \qquad (32)$$

$$-R_{12}' (H_1PH_1' + R_1)H_1PH_2'$$

where $\text{Cov}(V_1, V_2) = :R_{12}$.

The complete equations for sequential estimation in this case are the same as (27) and (28) with R_2 in (28) replaced by $R_2{}^n$ given in (32).

The error covariance after processing the first measurement is

$$P_1 = P - PH_1' (H_1PH_1' + R_1)^{-1} H_1P \qquad (33)$$

The error covariance after processing the second measurement is

$$P_2 = P_1 - P_1H_2' (H_2P_1H_2' + R_2{}^n)^{-1}H_2 P_1 \qquad (34)$$

It is important to note that although it was assumed in the development of (19) that the measurements themselves are available from the local estimate, this is not a requirement. It is only necessary that the required combination of the measurements can be recovered from the local estimate. In such a case the local estimates are said to be sufficiently informative.

The Utility of the Approach

It is important to note that with linear estimation techniques and linear measurement models as in (6) and (7), the equations of "sensor fusion" give the same results as with conventional Kalman filtering. There is one property which is unique to the sensor fusion idea, using feedback of the a priori state estimate in the local processor. This point may be easy to miss, since it cannot be explained with the usual filter formulation.

The linear measurement model

$$y_k = H_k x_k + v_k \qquad (35)$$

is used in linear filter theory. The a priori estimate $\bar{x}_k$ does not affect the measurement. However, in practice, this model may be incorrect. For example, in a range measuring device, the a priori estimate may be used to set the range and the speed gate. If the a priori estimation error is small, the target will be within the range gate and a good measurement will be made. If the gate setting is far off, the target may be completely missed. In this case the linear measurement model (35) does not describe the real situation well.

When the local estimators are non-linear, then the conventional Kalman filter does not lead to the same results. This is the situation where sensor fusion may give considerable improvement.

III.   SENSOR MODELS

It was pointed out that in the case of the linear measurement model, the results of sensor fusion are identical with the results which can be obtained with less work from the usual Kalman filter. Therefore, in order to demonstrate the advantages of sensor fusion in this application, a non-linear local processor was required. For illustration, we have selected a strapdown active RF sensor with a non-linear local processor. Another feature which must be present to demonstrate sensor fusion is a second aiding sensor; for that one, we have selected a strapdown imaging infrared (IIR) sensor.

The IIR seeker provided LOS angle information and was linear except for a field of view (FOV) limitation. Since this model was linear, no detailed non-linear local estimation for the IIR was used. The simulated RF sensor provides LOS angle measurements as well as optional target range and range rate information. Simple linear models were used for the range and range-rate measurements. The RF angle measurement were simulated by including a more detailed model which included the antenna pattern. Therefore, a non-linear local processor for the RF angle measurements was required.

In order to easily change the active sensors, sequential measurement updating was used. If all of the sensors are working as expected, sequential processing should improve the accuracy over batch processing for the extended Kalman filter as the a priori estimate which is used in the linearization is improved as each measurement is processed. It is important to note that although a number of the measurements were "linear", e.g., the measurement of an angle plus noise, they were non-linear functions of the states so the extended Kalman filter formulation was used.

The RF model used is based on a phased array monopulse system. The phased array consists of a number of individual radiators which are positioned and fed such that the resulting beam is highly steerable. Since the steering is by electronic switching which is very fast, a single array can track multiple targets or be switched in direction much faster than a parabolic reflector and a mechanical servo system.

For lack of space, the detailed derivation of the non-linear maximum a posteriori estimator is not included here.

IV.   FAILURE DETECTION

One of the major problems with the use of data from multiple sensors is the failure to detect bad measurements. Fortunately, the filter (pseudo) innovations sequence, available for each sensor, provides a means of detecting failed sensors or sensors which are not operating as expected. In this study, statistical decision theory was used to examine the hypothesis that all sensors were performing as predicted. Sensor failures were detected and bad measurements were discarded. The local estimators were continued and restarted if it was decided that they were again deemed satisfactory. A number of failure modes were considered including

loss of measurements, step increases in measurement noise covariance, and shifts in the measurement noise mean.

V.   SIMULATION STUDY

In order to demonstrate the performance of sensor fusion, the technique was studied using simulation. The University of Florida (UF) Bank to Turn (BTT) simulation program was modified to include a more detailed non-linear radar model and various noise environments. The equations for the missile dynamics simulated are for a generic bank to turn missile with a capability of 100 g acceleration in the pitch plane and 5 g's in yaw. The UF BTT simulation is a full six degree of freedom simulation which includes a choice of several guidance laws, estimation methods, and seekers used for guidance. A dither adaptive autopilot was also included in the simulation. Other features of the simulation which were necessary in order to study the possible value of sensor fusion are a non-linear sensor/local processor, a dual tracking sensor, and some type of sensor noise or a failed sensor.

Since the object of the simulation study was to examine the performance of the estimation schemes and not the guidance laws, the guidance law was fixed. The guidance law used is based on application of LQG theory to a nine state model for the relative missile target dynamics and is known as the AGL (advanced guidance law) in some of the literature. In all results reported here, the AGL was used with the target initially 40 degrees left off of the missile boresight. The initial range was 4000 feet with the missile and target both flying in the X-Y plane at 9900 feet. The target initially is headed toward the missile with a velocity of 989.9 feet per second and initiates a 10 g turn to the right. In order for the missile to hit the target, it must turn toward the target to the left. The missile must bank 90 degrees to place the maneuver plane in the missile pitch plane so that the full 100 g normal acceleration can be applied. A typical maneuver of the target and missile in the X-Y plane is shown in Figure 1. The same trajectory is shown in the X-Z plane in Figure 2.

The trajectories shown in Figures 1 and 2 show a hit on the scale shown, when sensor fusion was used. The same situation without sensor fusion is shown in Figures 3 and 4 where the target was missed by 1555 feet. In these simulations elevation, azimuth, range and range rate information were obtained from the RF Tracker. Angle information was also obtained from an IR sensor. The IR sensor had a +0.3 radian field of view (FOV) limit. The increase in the RF sensors noise environment was simulated by increasing the noise covariance in the angle measurements by ten sigma. This event, which was not detected, was active from 0.6 to 1.2 seconds.

Demonstration of Performance Improvement with Sensor Fusion

The failure of the missile with the increased noise and the operation of the estimation may be further studied by looking at the filter performance through plots of the filter innovations sequences. The innovations sequence is found by comparing the sensor outputs with the predicted sensor outputs. Figure 5 shows the innovations sequence for the IR angles and the corresponding predicted one sigma error bound with and without sensor fusion (SF). The target is only within the field of view in this case for one point.

The radar elevation and azimuth errors and predicted one sigma bounds for the same trajectory are shown in Figure 6 with and without SF. During the first part of the trajectory, the errors are within the error bounds and the filter is performing normally. Shortly after entering the noise environment at 0.6 seconds, the filter diverges. After the noise environment is terminated at 1.2 seconds, the errors are reduced, but remain far outside of the predicted values.

When the target comes into the field of view at 0.9 seconds with SF, the highly accurate IR measurement allows better pointing of the radar, in spite of the noise, so that the tracking errors are reduced. This allows the target to stay within the field of view for a longer time, further reducing the tracking errors. Finally, at 1.5 seconds, the target again leaves the IR FOV due to maneuvering. However, the RF noisy environment has terminated and the (RF sensor) can easily continue to track unaided until the very end of the encounter, where the relative motion becomes too fast.

Further information about the filter operation can be seem from Figures 7 and 8 which show the radar range and range rate errors with and without sensor fusion. During the countermeasure period (0.6 to 1.2 seconds) both range and range rate diverge. When sensor fusion is used, the errors recover and are again within limits when the noise environment is removed.

VI   CONCLUSIONS

Based on this study, the concept of sensor fusion has promise in making significant improvement in many systems. It is important to observe that good a priori estimates are necessary for non-linear estimation and these estimates can be obtained by combining information from several sensors. Also, sensor performance must be monitored and sensor failures detected to avoid bad estimates when using bad data.

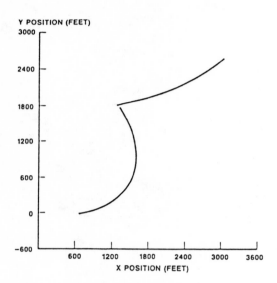

Figure 1   Target and Missile Trajectories
with Sensor Fusion - Plan View

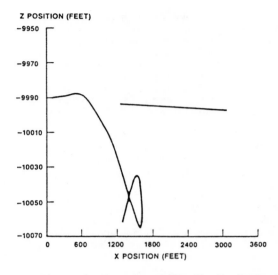

Figure 2   Target and Missile Trajectories
with Sensor Fusion - Vertical View

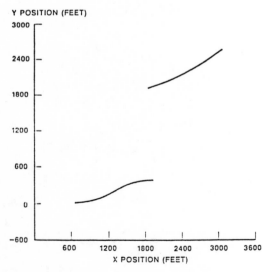

Figure 3   Target and Missile Trajectories
with no Sensor Fusion - Plan View

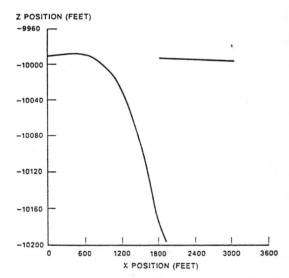

Figure 4   Target and Missile Trajectories
with no Sensor Fusion -
Vertical View

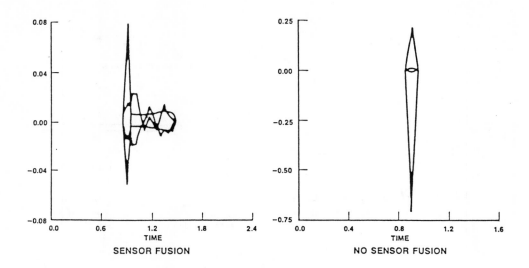

Figure 5   IR Angle Sensor Tracking Errors

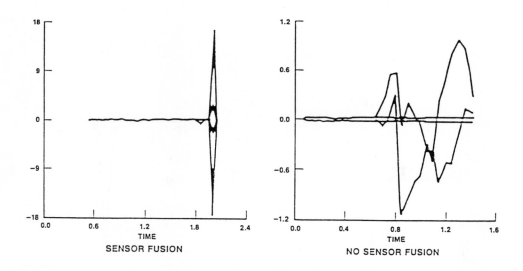

Figure 6   RF Angle Sensor Tracking Errors

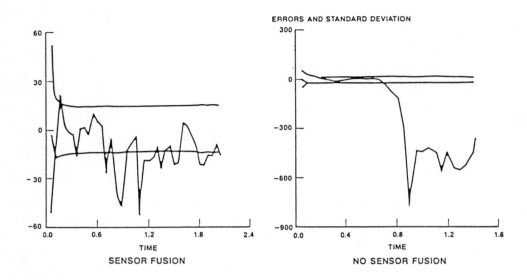

Figure 7   RF Sensor Range Errors

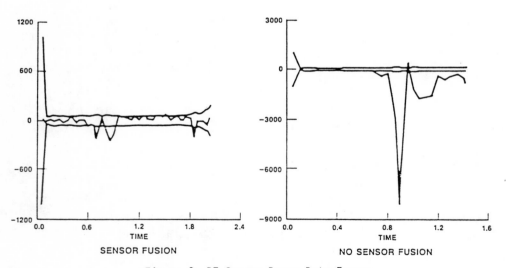

Figure 8   RF Sensor Range Rate Errors

A METHOD TO IMPROVE THE LOCATING ACCURACY
OF A MULTI-DETECTOR SYSTEM

Wenyi Chen

Optics Dept., Shandong University.
Jinan, Shandong, China

ABSTRACT

A method to improve the locating accuracy of a multi-detector system was proposed. In order to appraise the method, the target, the corresponding irradiance distribution function and the whole process for information extraction were simulated by the computer. The result shows that in some circumstance the locating accuracy will be improved in a great deal.

1. INTRODUCTION

A distant object may be considered as a point target; and most of the non-natural objects (including cooperative and non-cooperative) are symmetrical in shape. Although point targets and symmetric targets are not the whole entities interested by the infrared systems, but they still constitute the main part. For a multi-detector system, no matter scanning or staring, the accuracy of locating the objects is determined by the instaneous viewfield correspoding a single detector of it. In most case it is difficult to get more accurate position information then one pixel.

Is there any possibility to improve the accuracy without increasing the optical focal length and reducing the dimensions of each detector?

This problem is mainly concerned by this article. The image of a symmetric target is a distributional one, and the image of a point target on a plane of out-focus is distributional also. Commonly, they are symmetric in shape, and the symmetry could be used to improve the locating accuracy. A not too complicated method will be proposed in the following, and the increase of the accuracy will be proved by the computer simulation.

2. THE IMAGING GEOMETRY OF A STARING SYSTEM

Fig 1 gives the whole scheme , in which o'x'y': the object plane.

      oxy : the image plane, where the array-detecter is placed.

    O: a symmetric target

    I: the image.

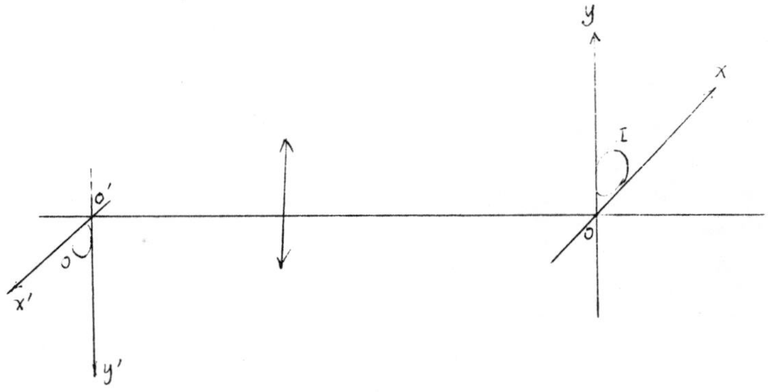

For a system, if the angle spanned by a distant object is less than the one spanned by a single detector, then the object will be considered as a point target, and the whole image of it will cover no more than one detecter. But the image will become distributional and covers several detectors if the array-detector is situated on the plane out-of-the-focus, or if the

Fig 1. The imaging geometry

effect of the diffraction will be considered. In the last case the distribution is symmetric for a perfect optical system, which is free of aberration , the irradiance distribution in the image plane is proportional to the square of first-order Bessel function. (ref.2)

3. THE OUTPUT SIGNAL OF THE DETECTOR

3.1. Supposition

The discuss will be conducted under the following ideal conditions:

3.1.1. The detectors of the array are the same, either in geometry or in optical and electrical properties.

3.1.2. The intrinsic noise of the system is very small and it works under the limit of the background noise.

3.1.3. The responsivity of the detectors is constant on the whole responsive plane.

3.1.4. The nearer to the centre of the image of the target, the larger of the irradiance.

3.2. The deductions:

3.2.1. The output signal of each detector is represented by an integral

$$S = \iint R(X,Y)I(X,Y)dxdy \qquad (1)$$

Because of the R(X,Y) is constant on the whole responsive plane and does not depend to x and Y, so it can be moved out of the integral, and we have

$$S=R\iint I(X,Y)dxdy \qquad (2)$$

It means the output is proportional to the flux of the whole detdctor. In the following the constant factor R will be removed for simplisity. For one dimension formula (2) will be reduced to : (see appendix)

$$S = \int I(x)dx \qquad (3)$$

3.2.2. The problem to locate the target is equivalent to the one of finding where is the greatest irradiance of the image.

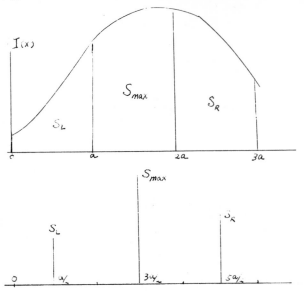

Fig 2. 1-d integrals of the signal

Considering Fig 2, the detector situated between the coordinats a and 2a; the output signal of it is the largest, then the signal will be expressed by

$$Smax = \int_{a}^{2a} I(x.y)dx \qquad (4)$$

and the signals of the left and right neighbor detectors will be

$$Sl= \int_{0}^{a} I(x.y)dx \qquad (5)$$

$$Sr= \int_{2a}^{3a} I(x.y)dx \qquad (6)$$

In which I(x) is unknow irradiance distribution function; Smax, Sl, and Sr can be measured in experiment. The problem which must be resolved becomes to decide the point M where the distribution function has the largest value.

4. LINEAR INTERPOLATION

For present multi-detector staring system, the location of a target is decided by the position of the detector which has the largest output signal. If the dimension of a single detector is 'a', then the largest possible deviation is 'a/2', the average absolute value of deviation is a/4 and the root-mean square deviation is 'a/2√3' (Appendix 2). The dimension 'a' limits the locating accuracy directly.

But if the deta are processed according to the flow chart given by Fig 3, then the accuracy will be improved greatly.

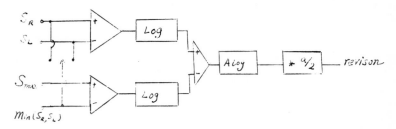

Fig 3. The flow chart of the linear interpolation.

In this figure `LOG` represents the logarithmic transformation; `ALOG` represents anti-LOG. The flow chart is designed according to the following physical considerations:

4.1. Smax is the largest output signal, the coordinate of the corresponding detector centre is epual to '1.5a'; Sl and Sr are the outputs of its left and right neighbor detectors. By linear interpolation of Smax and Sl (if Sl<Sr) we can find a point 'E', which has the same output as Sr (and vice versa when Sl>Sr). (see Fig 4). According to the geometry we can get:

OE=1.5a-x   (7)
and
    x:a=(Smax-Sr):(Smax-Sl)   (8)

4.2. The middle between 'R' and 'E' is the symmetric centre 'M' and has the largest irradiance of the image. So we have

OM=(OE+OR)/2
=1.5a +(Sr-Sl)*a/((Smax-Min(Sl,Sr))*2)
                                (9)

this formula is the fundation of the flow chart given by Fig 3, in which 1.5a is the result measured by present system with simple data processing method, and the remaining is the revision by linear interpolation.

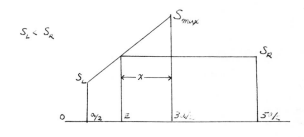

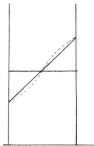

Fig 4. Linear interpolation

5. THE COMPUTER SIMULATION OF THE SYSTEM

By computer simulation the performance of the revision system could be appraised. The main contents are:

5.1. Simulation of the image
     irradance. There are 5 cases:
5.1.1. Linear

    I(x)=150-abs(x-xm)/2   (10)

5.1.2. Parabolic

    I(x)=150-(x-xm)(x-xm)/400
                                (11)

5.1.3. Exponent

    I(x)=150*exp(-abs(x-xm)/182)
                                (12)

5.1.4. Gause

    I(x)=150*exp(-(x-xm))/36140)
                                (13)

5.1.5. Sinc

I(x)=13155*sin((x-xm)/87.7)/(x-xm)
                                (14)

These functions are all symmetric. The centre of the symmetry has the largest value. Its coordinate 'xm' has value inbetween a and 2a , and a=50

5.2. Calculation of the output of the system.

According to the data flow chart compute the location of the centre, and compare it with xm

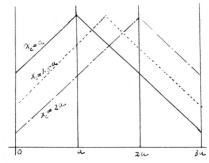

Linear fuction

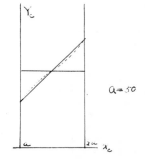

The maximum error=1.54
Mean absolute error=.71
The rms of error=.85

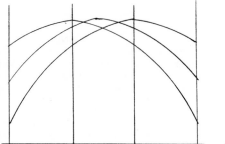

Quandric function

Maximum error=4.8
Mean absolute error=2.9
The rms of error=3.2

Fig 5. Computer simulation and the error.

value to get the differences.

5.3. Calculation of the error.

Supposing the probabilities of emergence of the target are equal at each point, compute the maximum deviation, the average absolute deviation , and the rms of the deviation. The results are showed on the right part of Fig 5, and it shows that the performance of the revision system is quite good.

5.4. The captions of Fig 5.

Xc--the location of the simulated irradiance centre.
Yc--the computed location of the centre.
The diagonal real line--the accurate location of the centre.
The horizontal real line--the resulted location from the present system.
The dot line--the resulted location from the revison system.

6.DISCUSSION: THE POSSIBILITY TO ENLARGE THE APPLICATIONS

If the situations are : neither the image of the target is symmetric nor the irradiance at the centre of the target image is the largest. In these case to applicate this system directly will deduce some error. Any system can not be universal, it must be designed for a special purpose. If the size, shape, and the distribution of the irradiance of the image are known before hand, and if the digital technique is in use, then the revision could not be limited linearly and with only direct neighbors, and the system will produce more accurate results.

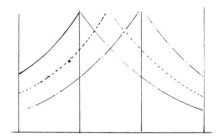
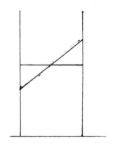

Exponent function

The maximum error=1.0
Mean absolute error=.50
The rms of error=.59

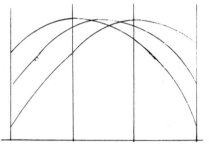
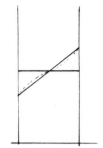

Gause function

The maximum error=4.5
Mean absolute error=2.7
The rms of error=2.9

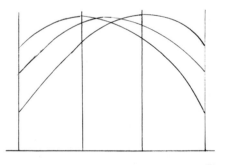
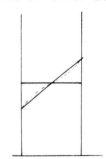

Sinc function

The maximum error=4.6
Mean absolute error=2.7
The rms of error=3.1

Fig 5. (continued)

7.REFERENCES

1. John A. Jamleson , Infrared technology: advance 1975-84, challenges 1985-94, Optical engineering, Vol.25, May 1986
2. W.L.Wolfe & G J Zissis(eds.), The infrared handbook,8-27,ERIM, 1978

APPENDIX

1. Reduction of the 2-d problem to the 1-d problem.
Oxy is the focus plane of the system. The configuration of the detectors in the focal plane array are shown in Fig 6. When working the system will sample the output signal of the detectors systematically. By comparison to find the largest output signal Smax first, and then the outputs of the four direct neighbors 'Su,Sd,Sl,Sr'. If the revision in x-direction is concerned, then the signals Smax, Sl,Sr will be processed; if the revision in y-direction is concerned, the Su, Sd,Smax will be processed instead.
2. The errors for present staring system.
the maximum deviation=a/2
the average absolute value of the deviation=$2\int_0^{a/2} x dx /a = a/4$
the rms of deviation=$\sqrt{((\int x^2 dx)/a)} = a/2\sqrt{3}$

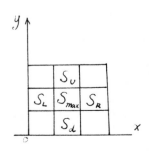

Fig 6. 2-d problem reduction

Target Acquisition and Track in the Laser Docking Sensor

by Ted J. Clowes and Richard F. Schuma

of Cubic Defense Systems Inc. and Cubic Electro-Optical
9333 Balboa Ave, San Diego, CA 92123 ; 750 Huyler St., Teterboro, NJ 07608

ABSTRACT

A sensor designed to aid in the docking of spacecraft is under development for NASA by Cubic Corporation. This sensor uses three lasers to track the prospective target and to determine the required parameters necessary to calculate the ideal approach maneuver. The system combines the inputs from several sensors, including polarization, continuous tone DME, and a CID to achieve the desired results.

1. INTRODUCTION

The laser docking sensor (LDS) is for use by the space shuttle to acquire and track satellites during recovery. Initially, the outputs; range, range rate, bearing (azimuth, elevation), attitude (roll, pitch, yaw), and attitude rate of the target will assist in manual approach and docking. After successful demonstration, the sensor can be integrated into the flight controls for automatic docking manuevers. Satellites will be acquired and tracked over a 20° by 20° field of view (FOV) at ranges from 0.1' to 22,000'. Attitude of the satellite will be determined by differential ranging and polarization of the return signal. Attitude rate will be determined by differential range rate.

The approach includes three semiconductor laser transmitter/receiver subsystems and a passive retroreflective target assembly. The three lasers will operate at different wavelengths. Narrow band filters, dichroic beamsplitters, and low scatter optics will be used to optically isolate the subsystems and prevent crosstalk. The subsystems will be located in the shuttle bay within a volume of 16" by 16" by 6.75". The passive target assembly, as shown in figure 1, will be located on the docking face of the satellite. The coordinate system for the LDS is shown in figure 2.

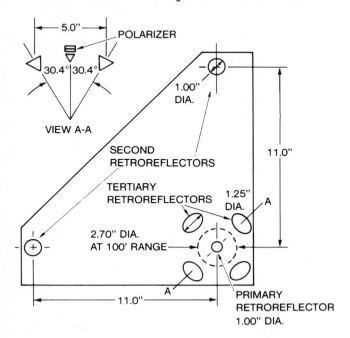

Figure 1. Satellite Target Assembly

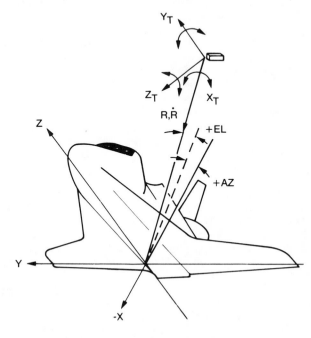

Figure 2. LDS Coordinate System

The three active subsystems are: (1) a long range bearing sensor (LRBS) which includes a pulsed semiconductor laser and a charge injection device (CID) camera; (2) a short range bearing/roll sensor (SRBS) which includes a CW laser, a lateral effect detector, two Wollaston prisms, and four photodiodes; and (3) a distance measuring equipment (DME) sensor which includes a CW laser, fiber optic links, and three avalanche photo diodes (APD). There will also be three sets of scan mirrors. Associated with one set of mirrors is an optical encoder mirror position sensor that uses two light emitting diodes (LED), two bi-cell detectors, two quad-cell detectors and two galvanometric single axis mirror controls. All subsystems are controlled by two microprocessors communicating through a shared memory. The processors are responsible for control and coordination of the different active components, as well as, melding the results from the sensors to provide the desired outputs. All three subsystems will operate through a primary port in the LDS. Two additional secondary ports will be used for differential ranging to measure pitch and yaw.

2. LONG RANGE BEARING SENSOR (LRBS)

The long range bearing sensor includes a CID camera with 512 x 512 pixels and a pulsed semiconductor laser illuminator operating at a wavelength of 797 nm. The LRBS is used at ranges between 22,000' and 100' for target acquisition and bearing relative to the LDS line of sight. At ranges less than 100', defocusing in the LRBS necessitates use of the SRBS for bearing measurements. The LRBS operates through the primary port of the LDS. Precise bearing measurements are made by combining the output of the CID camera and the mirror position sensors.

An anomorphic lens shapes the LRBS pulsed laser beam to match the active FOV of the CID camera, which is 0.117° horizontal by 3.0° vertical. The pulse rate of the laser is synchronized to the frame rate of the camera (165/second). The energy per pulse is controlled by varying the duration of the pulse, with a counter from 70ns to 4ms. While the pulse is on, the scan mirrors will be stabilized to prevent image smearing. The parameters that govern the satisfactory operation of the LRBS are the flash rate, flash duration, field of view, and the mirror/target motion. To meet the required acquisition time (30 seconds) with a moving target (1°/second), the mirror sweep rate needs to be 18°/second. This assumes a 3° x 0.117° FOV, a flash rate of 165/second, and a 50% overlap raster scan.[1]

While the CID array actually has 512 x 512 pixels, the active portion, as determined by the beam width of the laser will be 20 x 512 pixels. Another contiguous 20 x 512 pixel band will be used passively as a means of discriminating between retroreflective and stellar targets. This means the majority of the camera is unused. During calibration, the camera is checked to find at least 50 contiguous rows with no dead pixels. Each pixel is like a photodiode by itself and is measured for its incident radiation to distinguish shades of gray. Since subpixel averaging will be required to determine the correct bearing, the CID is required to operate near the middle of its dynamic range at all times. The CID can be directed to fast scan to a specific row before sampling a band of data. This allows control of the scanning and background windows, as well as, the amount of overshoot of the laser beam. Discrete control of the camera handles clearing pixels, sample start/stop, etc. It is also possible to load the horizontal and vertical count registers with the starting pixel index for this frame. This allows absolute pixel indices to be used for the entire 20° x 20° acquisition FOV. The address counter which is resettable and readable controls the location of sampled pixel data. Threshold control selects the values of pixel data that are actually stored in the pixel memory.

For each frame of camera operation the horizontal and vertical indices are set to reflect the current frame's upper left hand mirror position. Once started, the pixel value is compared with the threshold. If it exceeds threshold, the seven bit flash converted analog value is stored along with the primary port DME lock state. At the same time the horizontal and vertical pixel addresses are saved. The address counter control automatically steps to the next available location. To ensure the central processor can access pixel memory without dropping CID data, a set of FIFOs receive the actual data from the CID and index counters before it is stored in memory. The organization of memory is: 16 bits allocated to the "x" coordinate index of which the most significant 4 are non-existent; 16 bits allocated to the "y" coordinate index of which the most significant 4 are non-existent; 16 bits allocated to pixel value, the upper 8 are non-existent and the most significant bit in the low byte is DME lock; and 16 bits allocated to a repeat of the pixel value word to keep the addressing easy. All of these locations can be accessed as regular memory. The address counter generates an interrupt to the processor every time it rolls over (32768 samples over threshold). The starting value is preset to the maximum value (32767) since the counter increments before storing. In this fashion, it is also possible to tell when the first pixel exceeds threshold.

3. SHORT RANGE BEARING SENSOR (SRBS)

The SRBS performs the following functions: target bearing measurement form 0.1 to 100 feet; and roll attitude measurement. A single 20 mW CW semiconductor laser operating at 860nm is projected through beam shaping optics and a quarter wave plate to form a circularly polarized beam with a divergence angle of 0.117°. A linear polarizer positioned in front of the primary retroreflector in the target assembly converts the incident circular polarized light to a linear polarized retroreflected beam. After the return beam passes through a narrow band optical filter in the SRBS, it is split into two parts, one for the short range bearing calculation and the other for the roll measurement.

Bearing measurement is made by a lateral effect cell. This device measures the centroid of the incident radiation by creating a current flow to each of four contacts. Assuming A is the right, C is the left, B is the top, and D is the bottom contact, the centroid can then be determined by

$$x=(A-C)/(A+C) \qquad (1)$$
$$y=(B-D)/(B+D) \qquad (2)$$

For roll the incoming signal passes through a pair of Wollaston Prisms each of which produces two orthogonal components of the polarized signal. The two sets of signals are in quadrature so that maximum sensitivity can be achieved for all roll orientations. These signals are projected onto four photodiodes that have their outputs converted to digital values. The measurement of the roll angle is done by calculating θ. With A_i as one-half the range of the APD output and $(1+C_i)$ as the associated midpoint, the outputs from the four sensors are:

$$I_1=A_1(1+\sin 2\theta)+C_1 \qquad (3)$$
$$I_2=A_2(1+\cos 2\theta)+C_2 \qquad (4)$$
$$I_3=A_3(1-\sin 2\theta)+C_3 \qquad (5)$$
$$I_4=A_4(1-\cos 2\theta)+C_4 \qquad (6)$$

An ambiguity, of 180° still remains that requires external resolution by the use of the secondary ports. Roll can be tracked from either the tangent or the cotangent output by just determining the quadrant of the function through magnitude checks. This allows the function with the most linearity to be used.

4. DISTANCE MEASURING EQUIPMENT(DME)

The optical portion of the DME is shown in figure 3. The DME uses all three ports within 100' of range or the primary port only beyond 100'. A CW laser at 830nm is the source for the DME measurements. The laser power output is modulated by the electronic DME. This AM signal is used in lieu of the more conventional FM signal to generate continuous tones. These tones provide a measure of distance by detecting a phase shift from the reference signal.

Since the DME must operate over a large range of signal levels, a number of adjustments of the signal strength must be made for correct operation. These adjustments are made to keep the signal within the dynamic range of the APD by an optical attenuator. This device is a circular graded filter that has a continuous optical attenuation from 0 db to 80 db. It also has a wedge that is optically opaque for use during calibration or BIT. The wheel is driven by a motor to the desired position under software control based upon the APD signal level. After the signal has been attenuated, it is split into either one or three parts for the individual ports via fiber optic links. Beyond 100', all of the output power is sent to the primary port. Within 100', it will be equally partitioned to the three ports. The split is made by a combination of a 2x2 evanescent fiber optic coupler and a beamsplitter, as shown in figure 3.

Upon return, the signal passes through an optical filter and illuminates an APD. The AM modulated APD signal is then routed to the electronic portion of the DME, to measure the phase delays of the return from the reference transmitted signal by comparing a series of modulated tones. These phase delays are a measure of the distance traveled. The doppler frequency of the return signal is also compared to the reference for a range rate measure.

The LDS uses five ranging tones: fine at 374.74057 MHz; intermediate at 23.421286 MHz; coarse at 1.4638304 MHz; very coarse at 91.489398 KHz; and an extended at 11.436175 Khz. Each tone is a factor of 16 from its neighbors, except for the extended tone which is a factor of 8. This combination combined with a 11 bit phase meter provides a range of values from .1952mm to 13.1072km. Practical power limits prevent tracking to the extremes of these numbers. In practice, the tones are folded around the intermediate tone to make the signal processing easier. The intermediate and fine tones remain the same, but the transmitted tones are: intermediate plus coarse at 24.885116 MHz; intermediate plus very coarse at 23.512775 Mhz; and intermediate minus very coarse minus extended range at 23.318360 MHz. To reduce the associated hardware, only three tones are used at any one time. As a result, acquisition and target search use the three coarsest tones, and track uses the three finest tones. To further reduce the noise problems in the solution the tones are overlapped with each other in resolution, by the factors above. Each of the tones is phase measured with an 8 bit device, except for the fine and very coarse which share a 11 bit phase meter. This allows the final range measurement to be made up of a series of shifts and masks combining the various tones.

The range rate measurement is made by loading a counter with the expected number of wavelength transitions if there were zero doppler for the integration time (The counter is a modulus device that produces ambiguities in the measurement when a very high frequency is integrated for very long). If the dynamics of the target are bounded then these ambiguities can be ignored, as long as, the integration interval remains within the allowed target dynamic boundaries. When the counter is read at the end of the interval, the two's complement value indicates the number of bits of doppler shift that have occurred during the interval. The resolution is based upon the fine range tone frequency. The value (or velocity) of each count in the counter, with T_r being the integration time is equal to

$$v = 0.0244mm / T_r \qquad (7)$$

For example, if the integration time was 100ms, v=0.244mm/sec for a count of 1. To properly load the counter requires the computation of the expected count, where the constant 2 in the equation is the number of counts lost in the load circuitry, the 11.436KHz is the extended range tone frequency which is used as the reference, the 2^{14} is due to the multipliers in the RF, and the 2^{16} is the size of the counter, by

$$c = -\{(11.436 \text{ KHz} \times 2^{14} \times T_r / 2^{16}) -2\} \qquad (8)$$

For example, if T_r=100ms, then c=-284. It is also possible with this technique to compute an error difference of a doppler estimate*, if it was desirable to extend the tracking boundaries and an estimate of the doppler could be made (e.g. through first differences of the range measurements).

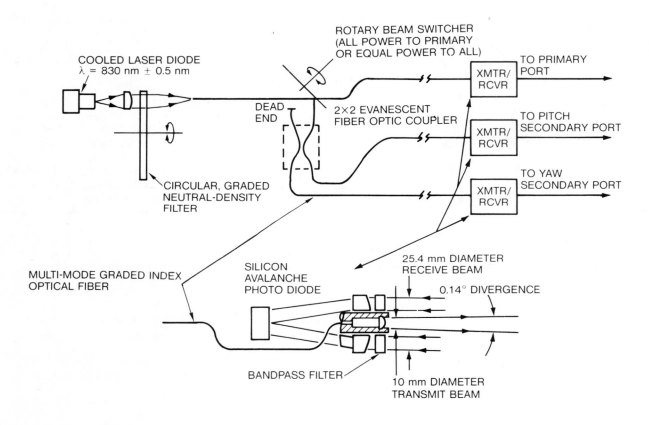

Figure 3. Optical portion of DME

5. MIRROR POSITIONING SENSORS

There are two mirrors and scanner motors associated with each port. The mirrors are for azimuth and elevation control of the outgoing laser beams. They are galvanometrically driven single axis devices capable of +/- 10° optical deflections. The primary port has mirror position sensors to assist the bearing sensors in determination of the

---

* The expected doppler, d= 2 x (v/c) x F_{fn}, is a number in hertz that must be added to the 11.436 KHz before it is multiplied by the other terms. For example a doppler of 6 m/s, with a 100ms integration time yields a load of -24,874.

bearing line to the target. The drives for the mirrors require 20ua/^{o}. Since the FOV of the bearing sensors is 0.117^o optical, the mirrors will be driven with as little as a 0.5ua signal change**. To achieve the required accuracy (0.0021^o), it is necessary to know where the mirror is pointing to 0.001^o. For this reason the primary port mirrors have an additional position sensing device. This device, shown in figure 4 , has two parts: a reference generating bi-cell and a quad cell that is used to count cycles and provide phase information.

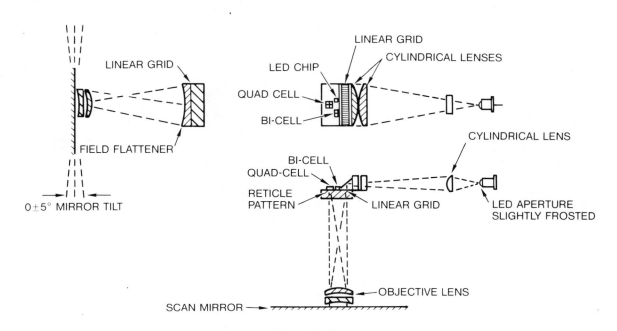

Figure 4. Mirror Positioning Sensor

A linear grid reticle with 1000 lines per inch is imaged via the back of the scan mirror onto a quad cell receiver. The quad-cell is covered by four grid patterns that cause the phase of the signal to be offset by 90° between quadrants to form a Lissajous pattern as the scan mirror is rotated. In this manner, it is possible to count the number of cycles to determine the mirror position. Each cycle is equivalent to 0.025^o. The outputs of the quad cell are evaluated to achieve the desired accuracy, using the same phase measurement equations (3-6) as used for the roll sensor. A zero index signal is generated when the image of a first pair of slits falls symmetrically on a second pair of slits to give a balanced signal from the bi-cell covering the second pair of slits.

6. TARGET ACQUISITION

The target acquisition process is entered when control is set to automatic operation either by data link command or by a control panel switch. It is assumed that nothing is known about the target position, other than the probability of existence within the 20° by 20° FOV. Acquisition is controlled by the central computer which clears all target data and starts the search, arbitrarily, in the upper left hand corner of the FOV. It is assumed the target is at maximum range, so the laser power is set assuming a return from a tertiary target retroreflector. The entire FOV is scanned and any spots that are detected are processed. During acquisition, each frame contains information about two slits of the FOV. The first slit is the illuminated area and the other is a non-illuminated area that will be illuminated in the future. The background routine uses the non-illuminated data to dynamically modify the pixel threshold comparator to help suppress stellar returns in the illuminated region. Power adjustments are made if non-illuminated targets provide too high an irradiance level. The scanning hardware associated with the CID camera is set up every frame time by an interrupt routine.

The initial frame is 6ms with overlaps of 1 pixel in the azimuth direction and 200 pixels in the elevation direction. This approach assures an acquisition in 30 seconds; with the initial indication in 15 seconds or less. The remainder of the time is used by the search routine for target lock on. Synchronization between the optical and the

---

** The mirror drive needs to be capable of one-half the desired optical angle (0.0558^o), thus requiring a mechanical angle of 0.0279^o. The assumption is this position will be accurate to one-half of a least significant bit (LSB).

central computers is performed using interrupts passed through a shared RAM. The optical frame routine waits for an interrupt to start it. Once started it sends an interrupt to the central computer to indicate the beginning of each frame. The interrupt routine in the central computer sets up the CID timers, pixel counters, and threshold values based upon a dynamic memory table that is modified by the background acquisition routine.

The optical computer uses the same routine for the search and track functions. It is driven by a parameter table in shared RAM, that can be set up for whatever mode is required by the central computer. This approach allows the optical computer software to be much simpler. It waits for the beginning of the frame, then it picks up the starting values, performs the initial conditions, and goes into a loop until a signal is received indicating the frame is complete. This loop simply reads the current mirror positions, converts to absolute pixels, and saves this value along with the rest of the analog values in shared memory. The LRBS laser is flashed, the next position is calculated, the flash is completed, the mirrors are driven to the next position, analog parameters are adjusted, and the frame is complete.

7. TARGET TRACK

Track is a primary function of the LDS. Once track is entered, the output data is valid for the range to the target (e.g. attitude data is valid only below 100'). The overall frame rate for track is different than for acquisition and search. It is slower for a number of reasons: more computations need to be made on each sample; data is not required at a rate higher than 10Hz; dark samples are taken to keep the target level known; and the speed of the tracked object does not warrant additional computational power. The optical computer is set to sample at 20Hz with flashes. The central computer dark samples the target space in between the flashes.

Upon entry to track, a target has been chosen that has a valid range. This may or may not be the correct target, however, it will be tracked until either it is lost or track is broken externally (i.e. by serial interface or switch command). Track sets up the basic optics parameters for the pointing angles, flash speed, and laser power, then enters a loop until loss of track. Inside the loop processing varies as a function of range, except for the reading of range, range-rate, and the updated pointing requirements. Beyond 100' range, the LRBS is used with the SRBS inoperative. Sub-pixel averaging is done to calculate the center-point bearing. Inside the 100' distance, the SRBS is used with the LRBS turned off. Short range bearing is calculated from the lateral effect detector, while roll is measured using the roll sensor. This calculation uses a arctangent table lookup with adjustment of the most significant bits of roll through the use of the secondary pointing angles. The roll sensor is ambiguous by 180°. The secondary targets are acquired and the return is used to calculate the pitch and yaw, using the arctangent table. Attitude rate is calculated using differential range rate. When the bearing numbers are available, regardless of distance, the bearing rate calculation can be made. These numbers are then used to predict the pointing requirements for the next frame. Range-rate takes a minimum of 85ms to settle; therefore the measurement started at the beginning of the interval can be read near the end. After all values have been read they are formatted and output over the serial links.

8. SUMMARY

Through the use of multiple sensors, it is possible to derive results that each of the sensors by themselves could not achieve. The techniques required to combine the sensors require some form of programmable control that can be used to close what would otherwise be open loops. More work is to be done to determine how tight the control needs to be and the amount of information that can be gleaned. It may be possible to extend the operating conditions of the LDS at reduced accuracies by using information from the range to help in higher range rate cases, or the information from the CID to help the DME for longer range, or some other set of combinations.

9. ACKNOWLEDGEMENTS

The authors wish to thank the design teams at Cubic Defense and Cubic Electro-Optical for their support and helpful discussions. This paper describes work performed under NASA contract NAS9-17846 by Cubic Corporation.

10. REFERENCES

1. Laser Docking Sensor Tradeoffs, by Pete Nelson, Richard Schuma, Ted Clowes of Cubic Corporation; 12 January 1988; San Diego, Ca.; TM228-1. pp 12-16.

A NEURAL NETWORK ARCHITECTURE
FOR EVIDENCE COMBINATION

by

Ted Pawlicki

Department of Computer Science
State University of New York at Buffalo
226 Bell Hall
Buffalo, N.Y. 14260
Ph: (716) 636-3197|3180
email: pawlicki@cs.buffalo.edu

ABSTRACT

Neural network models have attracted the attention of many researchers in the pattern recognition domain. These models possess many interesting computational properties which include content addressable memory, automatic generalization, and the ability to modify their processing (learn) based on their input data. They also promise extremely fast implementations if they can be realized in special purpose hardware. Such special purpose implementations, due to the limits of integration, imply a finite number of *neurones* for any one system[2]. Under such constraints, the construction of large neural network systems implies parallelism among sub-modules. This paper presents an architecture based on fusing the outputs of several independent neural network systems in order to define a single aggregate system.

The system presented here recognizes handwritten ZIP code digits taken from pieces of United States Postal Service (USPS) mail. The overall system is composed of several sub-modules, each of which could be realized in a neural network of reasonable size. Parts of the system have been shown to achieve up to 75% accuracy processing digits at the rate of about one digit per second (real time). Currently, the neural network paradigm on which the system is based is being simulated on a serial machine (a *Symbolics* 3600 series lisp machine). In order to keep the total time of the system within a reasonable limit of ~100 time steps, each *network module* has been limited to a few (<10) time steps. Current work involves the definition of other modules whose evidence will be combined with the described module. The gross system architecture is designed to integrate multiple evidence sources. The overall goal is to have both neural network and symbol processing paradigms in a single system.

1. INTRODUCTION

Recognition of handwritten digits is a classic problem in pattern recognition. Many approaches to this problem, from simple template matching to sophisticated knowledge based systems, have been shown to obtain a moderate level of success. Neural networks and their computational properties have attracted the interest of researchers in the area of machine perception by presenting an exciting, complementary alternative to symbolic processing paradigms. They hold with them the promise of exceedingly fast implementations coupled with flexibility through self-organization (learning). The approach described in this paper attempts fuse the output of several systems into a single output of high confidence by using the adaptive properties of neural networks.

The system operates on a data set consisting of 7756 16 by 16 binary arrays which represent numeric characters digitized from handwritten ZIP codes. In many cases the digits were touching or overlapping and had to be forcefully segmented. The ZIP code images were taken from a sample of live mail that was digitized at the Buffalo USPS Sectional Center Facility. The original images were scanned at 300 pixels per inch. The digit images were thresholded and size-normalized to a 16 by 16 grid for our experiments. The experiments used 1549 training and 6207 testing digits .

Figure **1** shows the context in which this module is intended to operate. For the purpose of recognition, the system is divided into a blackboard with a single, central *control network*. The central control network has two functions. It must integrate evidence from the various *knowledge modules* external to it. It must also internally make decisions about recognition taking place based on *a priori* knowledge as well as evidence combination.

In order maintain the practicality of the system sketched in figure **1** it is necessary to enforce several constraints upon its definition. These constraints are aimed at limiting the system to a tractable size while achieving reasonable throughput times. Current design constraints are as follows:

(1)   Each module (including both control and knowledge modules) should be limited to a reasonable number of *neurones*. The largest implemented to date contains 522 nodes and 5,220 connections.

(2)   Each module must independently converge on an output solution in less than ten logical time steps. This requirement is based on what has been called *the 100 step rule* motivated by insight gained from biological systems coupled with the heuristic that hierarchies of systems may be defined which are no more than ten deep. Currently defined modules each require only one logical time step and the current hierarchical structure is only three layers deep.

(3)   While dense interconnection within a single module is permitted. Inter-module communication must be limited. Currently, the largest amount of inter-module communication consists of ten logical lines each of which corresponds to a confidence level for a distinct digit zero through nine. This limit is specific to the current application. More complex applications might use more sophisticated coding

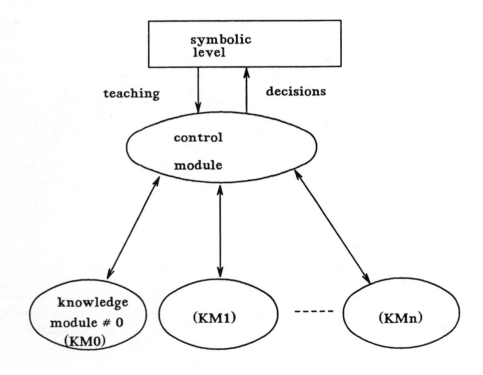

Figure 1

The Gross system architecture

techniques in order to transfer confidence levels with limited lines.

1.1. A SIMPLE EXAMPLE

Figure 2 shows a scaled down version of how data from two independent systems is merged into a single output. Input vectors V_1 and V_2 represent raw data from two independent sensor modalities. The input vector V_1, and the corresponding output vector O_1, along with the connecting network represent a single independent knowledge module, as do vectors V_2 and O_2 along with their connecting network. Networks 1 and 2 can be separately trained to give output in some canonical form.

The two output vectors O_1 and O_2 in turn act as input to the control module's network which is trained to classify the output of the knowledge modules. The effectiveness of this architecture is based not only on the ability of the knowledge modules to make correct analysis of the input data, but also on the ability of the upper level control module to adjust so as to correctly interpret the output of the knowledge modules. In the case where the knowledge modules make *consistent* mistakes the control module should be able to learn about these mistakes in order to ultimately arrive at a correct classification.

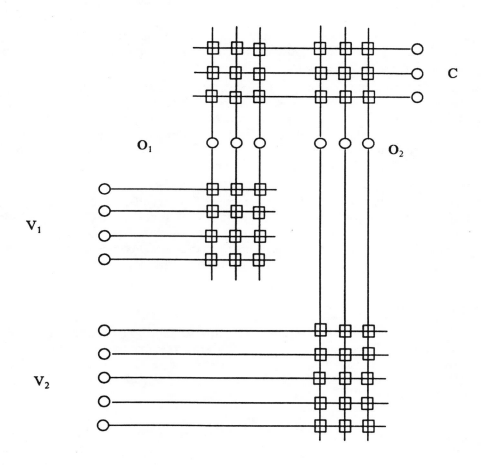

Figure 2

*An example network consisting of two
knowledge modules and one control module*

2. THE CURRENT SYSTEM

At present, two knowledge modules and a single control module have been defined. The focus of this paper, however, is on the methods used to combine evidence from arbitrary knowledge modules rather than on the precise representational issues of the knowledge modules. A brief discussion of the existing modules is included for clarity. A more detailed description of the existing system can be found in Pawlicki and Srihari (1988)[3].

The first knowledge module operates on a simple template matching scheme. The input consists of a 16 by 16 bit map image of a digitized, normalized, thresholded digit extracted from a hand-written zip code off a live piece of USPS mail. The 16 by 16 pixels are completely connected to ten output nodes which correspond to the digits zero through nine. Teaching input is used to train module independently. The bit mapped knowledge module can independently achieve up to 74% correct rates.

The second knowledge module which has been defined maps chain code representations of character shapes into a neural network vector. A chain code consists of a string of digits from zero to seven, which correspond to eight possible orientations of the shape contour at any one point. An 8 by 8 by 8 space which is subsequently mapped into a vector of length 512 is used to represent triples taken from the chain code string. This is an adaptation of Wickelgren's scheme[4] that was used by Rumelhart and McClelland (1987)[1] who applied a neural network paradigm in the domain of linguistics to learn the past tenses of English verbs. While some information is lost in this transformation, the knowledge module based on this representation is capable of up to 60% correct rates.

Neither of the two knowledge modules achieve correct rates which would be acceptable as a character recognition system in a real world application. The thrust of this research is not to define a single perfect character recognizer, but rather to develop a system capable of merging the imperfect outputs of several systems into a single output at a higher level in order to improve performance. At present a single control module operates by concatenating the output vectors of the two existing knowledge modules into a single input vector of length 20. The control module is independently trained not on the raw data, but on the output vectors of the two knowledge modules.

Experiments are presently under way to determine the ability of such a control module to learn about the behaviors of its input knowledge modules. Present results reveal that under some learning paradigms the control module is able to correctly classify a character in the case where one of the knowledge modules makes an incorrect classification and the other is correct. If this result can be maintained consistently, over a large data set, the control module will effectively be operating on the basis of the best of both its knowledge modules.

3. FUTURE DIRECTIONS

Current work involves experimentation with a number of different learning paradigms in order to train modules at all levels. Experiments aimed at assessing the maximum correct percentages realizable through this architecture are in progress. The control module is also being modified in order to accept confidence levels from the knowledge modules for the purpose of selectively weighting their relative importance.

References

1.   D. E. Rumelhart and J. McClelland, "On Learning the Past Tenses of English Verbs," in *Parallel Distributed Processing - Chapter 18*, ed. McClelland, (1987).

2.   J. J. Hopfield and D. W. Tank, "Collective Computation in Neuronlike Circuits," *Scientific American*, pp. 104-114 ().

3.   T. F. Pawlicki and S. N. Srihari, *Proceedings of the Indo-US Workshop on Systems and Signal Processing*, (1988).

4.   W. A. Wickelgren, "Context-sensitive coding, associative memory, and serial order in (speech) behavior," *Psychological Review* **76** pp. 1 - 15 (1969).

NETWORKING DELAY AND CHANNEL ERRORS IN DISTRIBUTED DECISION FUSION

Stelios C.A. Thomopoulos[1] and Lei Zhang
Department of Electrical Engineering
Southern Illinois University
Carbondale, Illinois 62901-6630

ABSTRACT

The effects of transmission delay and channel errors on the performance of a distributed fusion system is studied. At a given time instant, the decisions from some sensors may not be available at the fusion center due to the transmission delays. Assuming that the fusion center has to make a decision based on the data from the rest of the sensors, provided that at least one peripheral decision has been received, it is shown that the optimal decision rule that maximizes the probability of detection for fixed probability of false alarm at the fusion center is the Neyman-Pearson test at the fusion center and the sensors as well. Furthermore, it is shown that, in the case of noisy channels, the decision made by each sensor depends on the reliability of the corresponding transmission channel. Moreover, the probability of false alarm at the fusion is restricted by the channel errors. For a given decision rule, the probability of any channel being in error must be kept at a certain level in order to achieve a desired probability of false alarm at the fusion. A suboptimal, but very near-to-optimal, computationally efficient algorithm is developed to solve for the sensor and fusion thresholds sequentially. Numerical results are provided to demonstrate the closeness of the solutions obtained by the suboptimal algorithm to the optimal solutions.

SECTION 1. INTRODUCTION

Problems dealing with distributed decision fusion have been receiving a lot of attention in recent years [1-9]. Based on the Lagrange multipliers method of minimization, Srinivasan [10] has proved that the globally optimal solution to the fusion problem that maximizes the probability of detection for fixed probability of false alarm when sensors transmit independent, binary decisions to the fusion center, consists of Neyman-Pearson (NP) tests at all sensors, including the fusion. However, the optimal thresholds obtained from this method hold true only if the Lagrange multipliers method is valid. For certain decision rule, the Lagrange method may fail to give the optimal solution [11]. A general proof of the optimality of the N-P test for the distributed decision fusion problem with independent decisions, which does not depend on any specific minimization technique was obtained by Thomopoulos, Viswanathan, and Bougoulias [12-13]. In [12], a suboptimal sequential solution to the fusion problem in terms of a one-dimensional search algorithm was developed. In [12-13] it was assumed that the channel was errorless and that there were no delays in the decision transmission, i.e., all the decisions were present at the fusion at the time of fusing. In this paper we consider the distributed decision fusion problem when decisions are transmitted with delays resulting from networking and bandwidth constraints, over a noisy channel that introduces some probability of error.

Consider the distributed sensor fusion system shown in Figure 1. We assume that at the time of fusion t, the peripheral decision of the i-th sensor, u_i , is available at the fusion center with probability p_i (i=1,2,...,N). The sensor decisions are assumed to be binary, i.e.,

$$u_i = \begin{cases} 1, & \text{if the decision of the i-th sensor favors hypothesis } H_1 \\ 0, & \text{if the decision of the i-th sensor favors hypothesis } H_0 \end{cases}$$

The probability of u_i being received correctly by the fusion is $1-P_{c_i}$ (P_{c_i} is the probability of channel i being in error). The p_i's and P_{c_i}'s are considered as the networking parameters and are assumed to be statistically independent. The fusion center makes a decision at time t based on the data available at that time. In previous studies, we only considered the cases in which the peripheral decision from each sensor is always available at the fusion center at the time of fusion, and is always received correctly, which is a special case of the model to be discussed here (by setting $p_1 = p_2 = \ldots = p_N = 1$, $P_{c_1} = P_{c_2} = \ldots = P_{c_N} = 0$, the model used in [10] and [12-13] results). Our objective is to investigate how the networking parameters affect the performance of the distributed decision fusion system, and derive the optimal test that maximizes the probability of detection at the fusion for a fixed probability of false alarm.

The model that we consider captures the effect of the delays in the sensor network in the following way. Assume that the fusion center times out for a fixed time-interval between two consecutive fusions. During this fixed time-interval, decisions from the sensors arrive with probability p_i. Hence, p_i can be thought of

_____

1. This research is sponsored by the SDIO/IST and mamaged by the Office of Naval Research under Contract N00014-86-k-0515.

as the probability that the networking delay at the i-th sensor does not exceed the time-out interval at the fusion.

The paper is organized as follows. Section 2 contains the general definitions and notations used in later sections. In Section 3, 4 and 5, it is shown that the optimal test that maximizes the probability of detection at the fusion, for fixed probability of false alarm, is the Neyman-Pearson test at the sensors and the fusion in the case that either delay, or errors, or both are present. Furthermore, equations for the optimal set of thresholds analogous to those in [10] are obtained for all the three cases of delay, channel errors, or both. In Section 6, a suboptimal sequential algorithm is derived that allows the determination of vary near-to-the-optimal operating points. Numerical results are given in Section 7.

SECTION 2. DEFINITIONS AND NOTATIONS

For the model shown in Fig.1, the following notations are introduced to represent the decision sets received at the fusion center, the collections of such sets and their probabilities of appearance at the fusion center:

$U = \{ u_{i_1}, u_{i_2}, \ldots, u_{i_m} \}$, $1 \leq m \leq N$, $i_1, \ldots, i_m \epsilon \{ 1,2,\ldots,N \}$, and $i_j \neq i_l$ for $j \neq l$, is the set which contains at least one peripheral decision.

$U^k = \{ u_k, u_{i_1}, \ldots, u_{i_m} \}$, $0 \leq m \leq N-1$, $i_1, \ldots, i_m \epsilon \{ 1,\ldots,k-1,k+1,\ldots,N \}$, and $i_j \neq i_1$, for $j \neq l$, is the set which contains the decision from sensor k.

$U_k = \{ u_{i_1}, u_{i_2}, \ldots, u_{i_m} \}$, $1 \leq m \leq N$, $i_1, \ldots, i_m \epsilon \{ 1,2,\ldots,k-1,k+1,\ldots N \}$, and $i_j \neq i_1$, for $j \neq l$, is the set which does not contain the decision from sensor k.

$U_k^k = \{ u_{i_1}, u_{i_2}, \ldots, u_{i_m} \}$, $0 \leq m \leq N-1$, $i_1, \ldots, i_m \epsilon \{ 1,\ldots,k-1,k+1,\ldots,N \}$, and $i_j \neq i_1$, for $j \neq l$, is the remaining set of U^k after removing u_k from U^k, i.e., $U_k^k = U^k - \{u_k\}$.

S_k = the collection of all the sets which contain at least one peripheral decision.

S^k = the collection of all the sets which contain u_k.

S_k = the collection of all the sets which do not contain u_k.

$S_{k,k-1}$ = the collection of all the sets which contain both u_k and u_{k-1}.

S_k^{k-1} = the collection of all the sets which contain u_{k-1} but not u_k.

$P(U)$ = Prob { set U received by fusion center at time t | at least one of the peripheral decisions has been received }.

SECTION 3. OPTIMAL FUSION RULE WITH DELAYS AND IDEAL CHANNELS

Suppose that N sensors receive data from a common volume. Sensor k receives data r_k and generates the first stage binary decision u_k, k=1,2,...,N. During a fusion interval the decisions are transmitted with random delays through errorless channels and arrive at the fusion center with some probability where they are combined into a final decision u_0 about which one of the two hypotheses is true, where

$$u_k = \begin{cases} 1 : \text{decide hypothesis } H_1 \text{ is true} \\ 0 : \text{decide hypothesis } H_0 \text{ is true} \end{cases} \tag{1}$$

for k= 0,1,...,N. At the time the fusion makes a decision, we assume that at least one decision from a sensor has been received. According to our model, the fusion times-out after each decision for some fixed time interval. At the end of the time-out interval a new decision is made based on the decisions that arrived in this interval.

We assume that the data r_k received by sensor k is statistically independent from the data received by the other sensors conditioned on each hypothesis. Hence, the decisions u_k are statistically independent conditioned on each hypothesis, k = 1, 2, ..., N. Given a desired level of probability of false alarm at the fusion center, $P_F = \alpha_0$, we seek the optimal test that maximizes the probability of detection P_{D_0} (or minimizes the probability of miss $P_{M_0} = 1 - P_{D_0}$). In [14] we havv shown that the optimal test that maximizes the probability of detection for fixed probability of false alarm involves a Neyman-Pearson test at the fusion and Likelihood-Ratio tests at the sensors. The tests at the k-th sensor was shown to be

$$\frac{p(r_k|H_1)}{p(r_k|H_0)} \underset{u_k=0}{\overset{u_k=1}{\gtrless}} \lambda_0 \frac{C_0^k}{C_1^k} \tag{2}$$

where
λ_0 is the threshold at the fusion and

$$C_i^k = \sum_{U^k \epsilon S^k} \sum_{\{U_k^k\}} [d(u_k=1, U_k^k) - d(u_k=0, U_k^k)] p(U_k^k|H_i) P(U^k) \tag{3}$$

for i=0,1. From (3), it seen that the optimal set of thresholds depends on the network delays through $P(U^k)$, the probabilities of receiving a given set of sensor decisions.

Remark: The above derivation is valid provided that the optimal solution to the fusion problem does not lie in the boundaries of the decision space, in which case the solution is singular and the Lagrange multipliers method fails [12-13]. A general proof of the optimality of the N-P test could be obtained in a manner similar to [12] and [13]. The model that is used to incorporate the delay effect in the fusion problem is similar to the one used in [14].

SECTION 4. OPTIMAL FUSION RULE WITH NOISY CHANNELS

Next we consider the case in which the transmission delays are zero ($p_i=1$, for $i=1,2,\ldots,N$) but the channels are noisy. In this case, the decision from each sensor is always available at the fusion center, i.e. the set U received by fusion always contains N peripheral decisions. Let P_{c_i} be the probability that the i-th channel is in error, $i=1,2,\ldots,N$. The probability of false alarm P_{F_0} and the probability of miss P_{M_0} at the fusion center can be expressed as

$$P_{F_0} = \sum_{\{U\}} p(u_0=1 \mid U)\overline{p}(U \mid H_0) \tag{4}$$

$$P_{M_0} = \sum_{\{U\}} p(u_0=0 \mid U)\overline{p}(U \mid H_1) \tag{5}$$

where

$$\overline{p}(U \mid H_i) = \prod_{k=1}^{N} \overline{p}(u_k \mid H_i) = \prod_{k=1}^{N} [p(u_k \mid H_i, c_k)(1-P_{c_k}) + p(u_k \mid H_i, \overline{c}_k)P_{c_k}] = \prod_{k=1}^{N} [p(u_k \mid H_i)(1-P_{c_k}) + (1-p(u_k \mid H_i))P_{c_k}] \tag{6}$$

and

$$c_k = \{ \text{The decision } u_k \text{ is received correctly} \} \tag{7}$$

$$\overline{c}_k = \{ \text{The decision } u_k \text{ is received incorrectly} \} \tag{8}$$

$$p(u_k \mid H_i) = p(u_k \mid H_i, c_k) \tag{9}$$

Define the Lagrangian

$$F = P_{M_0} + \lambda_0(P_{F_0} - \alpha_0) = \sum_{\{U\}} p(u_0=1 \mid U)[\lambda_0 \overline{p}(U \mid H_0) - \overline{p}(U \mid H_1)] + 1 - \lambda_0\alpha_0 \tag{10}$$

The minimum value of F at the fusion is achieved if the decision rule

$$d(U) = p(u_0=1 \mid U) = \begin{cases} 1 & \text{if } \overline{p}(U \mid H_1) - \lambda_0\overline{p}(U \mid H_0) > 0 : \text{decide } H_1 \\ 0 & \text{if } \overline{p}(U \mid H_1) - \lambda_0\overline{p}(U \mid H_0) < 0 : \text{decide } H_0 \end{cases} \tag{11}$$

is chosen, or equivalently

$$\frac{\overline{p}(U \mid H_1)}{\overline{p}(U \mid H_0)} \underset{\substack{H_0 \\ u_0=0}}{\overset{\substack{u_0=1 \\ H_1}}{\gtrless}} \lambda_0 \tag{12}$$

Since

$$p(u_k \mid H_i) = \int_{r_k} [p(u_k \mid r_k)(1-P_{c_k}) + (1-p(u_k \mid r_k))P_{c_k}] p(r_k \mid H_i) \tag{13}$$

where

$$p(u_k \mid r_k) = p(u_k \mid r_k, c_k) \tag{14}$$

the Lagrangian in (10) can be expanded as

$$F = \int_{r_k} p(u_k=1 \mid r_k)[(1-P_{c_k})-P_{c_k}][\lambda_0 C_0^k p(r_k \mid H_0) - C_1^k p(r_k \mid H_1)] + F_k \tag{15}$$

where

$$C_i^k = \sum_{\{U_k\}} [d(u_k=1, U_k) - d(u_k=0, U_k)]\overline{p}(U_k \mid H_i) \tag{16}$$

$$F_k = P_{c_k}[\lambda_0 C_0^k - C_1^k] + \sum_{\{U_k\}} d(u_k=0, U_k)[\lambda_0 \overline{p}(U_k \mid H_0) - \overline{p}(U_k \mid H_1)] + 1 - \lambda_0\alpha_0 \tag{17}$$

From (17) we see that F is minimized w.r.t. the decision rule of the k-th sensor, if

Case 1: For $(1-P_{c_k})-P_{c_k} > 0$, or $P_{c_k} < 0.5$ and the decision rule

$$p(u_k=1 \mid r_k) = \begin{cases} 1 & \text{if } C_1^k p(r_k \mid H_1) - \lambda_0 C_0^k p(r_k \mid H_0) > 0 : \text{decide } H_1 \\ 0 & \text{if } C_1^k p(r_k \mid H_1) - \lambda_0 C_0^k p(r_k \mid H_0) < 0 : \text{decide } H_0 \end{cases} \tag{18}$$

is chosen, or equivalently

$$\frac{p(r_k \mid H_1)}{p(r_k \mid H_0)} \underset{\substack{\\ u_k=0}}{\overset{\substack{u_k=1 \\ }}{\gtrless}} \lambda_0 \frac{C_0^k}{C_1^k} \triangleq \lambda_k \qquad\qquad k=1,2,\ldots,N \tag{19}$$

Case 2: For $(1-P_{c_k})-P_{c_k} < 0$, or $P_{c_k} > 0.5$ and the decision rule

$$p(u_k=1 \mid r_k) = \begin{cases} 0 & \text{if } C_1^k p(r_k \mid H_1) - \lambda_0 C_0^k p(r_k \mid H_0) > 0 : \text{decide } H_1 \\ 1 & \text{if } C_1^k p(r_k \mid H_1) - \lambda_0 C_0^k p(r_k \mid H_0) < 0 : \text{decide } H_0 \end{cases} \tag{20}$$

is chosen, or equivalently

$$\frac{p(r_k \mid H_1)}{p(r_k \mid H_0)} \underset{\substack{\\ u_k=0}}{\overset{\substack{u_k=1 \\ }}{\lessgtr}} \lambda_0 \frac{C_0^k}{C_1^k} = \lambda_k \qquad\qquad k=1,2,\ldots,N \tag{21}$$

Equations (18) through (21) indicate that the optimal decision rule at the sensor is depend on the corresponding channel error. There exists a switching point $P_{c_k} = 0.5$, if the probability of channel error is less than 0.5, the sensor will transmit the true decision to the fusion, if it is greater than 0.5, the sensor will transmit the complement of the decision instead.

SECTION 5. OPTIMAL FUSION RULE WITH DELAY AND NOISY CHANNELS

When both transmission delay and channel error are considered, the optimal decision rule at the fusion center is given by

$$\frac{\overline{p}(U|H_1)}{\overline{p}(U|H_0)} \underset{\underset{u_0=0}{H_0}}{\overset{\overset{u_0=1}{H_1}}{\gtrless}} \lambda_0 \quad , \quad \text{where} \tag{22}$$

$$\overline{p}(U|H_i) = \prod_{u_k \in U} [p(u_k|H_i)(1-P_{c_k}) + (1-p(u_k|H_i))P_{c_k}] \quad , \quad \text{and} \tag{23}$$

$$c_k = \{ \text{The decision } u_k \text{ is received correctly} \} \tag{24}$$

$$p(u_k|H_i) = p(u_k|H_i, c_k) \tag{25}$$

The local decision rule is:

$$\frac{p(r_k|H_1)}{p(r_k|H_0)} \underset{\underset{u_k=0}{}}{\overset{\overset{u_k=1}{}}{\gtrless}} \lambda_0 \frac{D_0^k}{D_1^k} = \lambda_k \qquad k=1,2,\ldots,N, \text{ if } P_{c_k} < 0.5, \text{ or} \tag{26}$$

$$\frac{p(r_k|H_1)}{p(r_k|H_0)} \underset{\underset{u_k=0}{}}{\overset{\overset{u_k=1}{}}{\lessgtr}} \lambda_0 \frac{D_0^k}{D_1^k} = \lambda_k \qquad k=1,2,\ldots,N, \text{ if } P_{c_k} > 0.5, \text{ where} \tag{27}$$

$$D_i^k = \sum_{U^k \in S^k} \sum_{\{U_k^k\}} [d(u_k=1,U_k^k) - d(u_k=0,U_k^k)]\overline{p}(U_k^k|H_i)P(U^k) \qquad \text{for } i=0,1 \tag{28}$$

SECTION 6. SUBOPTIMAL SOLUTION

In Eqs. (2), (19) and (26) the optimal set of thresholds are given in terms of a set of nonlinear coupled equations whose solution depends on the fusion policy which is unknown. Hence, with the exception of very few simple cases, the equations that determine the optimal thresholds as obtained by the Lagrangian approach cannot be solved. Hence, a multidimensional search over all possible operating points of the sensors and all fusion policies needs to be carried through, if the optimal solution is sought. This, however, can be either computationally tedious, or, even infeasible [12-13].

Hence, a computationally efficient suboptimal algorithm has been developed to solve for the thresholds sequentially by minimizing the residual terms in the Lagrangian instead of the entire F [12], [15]. For the case in which both delay and channel error are considered, similar results can be obtained by applying the same procedure as above.

Let 1, 2, ..., N be an arbitrary numbering of the N sensors. The suboptimal decision rule for this case at k-th sensor is given by:

$$\frac{p(r_k|H_1)}{p(r_k|H_0)} \underset{\underset{u_k=0}{}}{\overset{\overset{u_k=1}{}}{\gtrless}} \lambda_0 \frac{D_0^k}{D_1^k} = \lambda_k \quad , \text{ if } P_{c_k} < 0.5, \text{ or} \tag{29}$$

$$\frac{p(r_k|H_1)}{p(r_k|H_0)} \underset{\underset{u_k=0}{}}{\overset{\overset{u_k=1}{}}{\lessgtr}} \lambda_0 \frac{D_0^k}{D_1^k} = \lambda_k \quad , \text{ if } P_{c_k} > 0.5, \text{ where} \tag{30}$$

$$D_i^k = \sum_{U^k \in S^k} \sum_{\{U_k^k\}} [d(u_k=1,U_k^k) - d(u_k=0,U_k^k)]\overline{p}(U_k^k|H_i)P(U^k) \tag{31}$$

For (k-1)th sensor, we have

$$\frac{p(r_{k-1}|H_1)}{p(r_{k-1}|H_0)} \underset{\underset{u_{k-1}=0}{}}{\overset{\overset{u_{k-1}=1}{}}{\gtrless}} \lambda_0 \frac{D_0^{k,k-1} + D_0^{k-1}}{D_1^{k,k-1} + D_1^{k-1}} = \lambda_{k-1} \quad , \text{ if } P_{c_k} < 0.5, \text{ or} \tag{32}$$

$$\frac{p(r_{k-1}|H_1)}{p(r_{k-1}|H_0)} \underset{\underset{u_{k-1}=0}{}}{\overset{\overset{u_{k-1}=1}{}}{\lessgtr}} \lambda_0 \frac{D_0^{k,k-1} + D_0^{k-1}}{D_1^{k,k-1} + D_1^{k-1}} = \lambda_{k-1} \quad , \text{ if } P_{c_k} > 0.5, \text{ where} \tag{33}$$

$$D_i^{k,k-1} = \sum_{U^{k,k-1} \in S^{k,k-1}} \sum_{\{U_{k,k-1}^{k,k-1}\}} [d^k(u_k=0,u_{k-1}=1,U_{k,k-1}^{k,k-1})-d^k(u_k=0,u_{k-1}=0,U_{k,k-1}^{k,k-1})]\overline{p}(U_{k,k-1}^{k,k-1}|H_i)P(U^{k,k-1}) \text{ for } i=0,1$$

$$D_i^{k-1} = \sum_{U_k^{k-1} \epsilon S_k^{k-1}\{U_{k,k-1}^{k-1}\}} \sum [d(u_{k-1}=1,U_{k,k-1}^{k-1})-d(u_{k-1}=0,U_{k,k-1}^{k-1})]\overline{p}(U_{k,k-1}^{k-1}|H_i)P(U_k^{k-1}) \qquad \text{for } i=0,1 \ ,$$

and
$$d^k(U_k^{k,k-1}) = P_{c_k}d(u_k=1,U_k^{k,k-1}) + (1-P_{c_k})d(u_k=0,U_k^{k,k-1}) \qquad (34)$$

is the average decision rule over the channel error of the k-th sensor.

For the (k-2)th sensor, the threshold test is

$$\frac{p(r_{k-2}|H_1)}{p(r_{k-2}|H_0)} \overset{u_{k-2}=1}{\underset{u_{k-2}=0}{\gtrless}} \lambda_0 \frac{D_0^{k,k-1,k-2}+D_0^{k,k-2}+D_0^{k-1,k-2}+D_0^{k-2}}{D_1^{k,k-1,k-2}+D_1^{k,k-2}+D_1^{k-1,k-2}+D_1^{k-2}} = \lambda_{k-2} \qquad \text{if } P_{c_k} < 0.5, \text{ or} \quad (35)$$

$$\frac{p(r_{k-2}|H_1)}{p(r_{k-2}|H_0)} \overset{u_{k-2}=1}{\underset{u_{k-2}=0}{\lessgtr}} \lambda_0 \frac{D_0^{k,k-1,k-2}+D_0^{k,k-2}+D_0^{k-1,k-2}+D_0^{k-2}}{D_1^{k,k-1,k-2}+D_1^{k,k-2}+D_1^{k-1,k-2}+D_1^{k-2}} = \lambda_{k-2} \qquad \text{if } P_{c_k} > 0.5, \qquad (36)$$

where
$$d^{k,k-1}(U_{k,k-1}^{k,k-1,k-2}) = P_{c_{k-1}}d^k(u_{k-1}=1,U_{k,k-1}^{k,k-1,k-2}) + (1-P_{c_{k-1}}) d^k(u_{k-1}=0,U_{k,k-1}^{k,k-1,k-2}) \qquad (37)$$

Finally, for the last sensor, i.e. sensor 1, we have

$$\frac{p(r_1|H_1)}{p(r_1|H_0)} \overset{u_1=1}{\underset{u_1=0}{\gtrless}} \lambda_0 = \lambda_1 \qquad \text{if } P_{c_k} < 0.5, \text{ or} \qquad (38)$$

$$\frac{p(r_1|H_1)}{p(r_1|H_0)} \overset{u_1=1}{\underset{u_1=0}{\lessgtr}} \lambda_0 = \lambda_1 \qquad \text{if } P_{c_k} > 0.5 \qquad (39)$$

where λ_0 is the threshold at the fusion center.

SECTION 7. NUMERICAL RESULTS

Performance evaluation of the distributed fusion system of Figure 1 in a slowly fading Rayleigh channel has been done numerically for the case of three sensors (N=3). For numerical convenience and for the sake of the presentation clarity, it was assumed that all of three sensors incurred the same networking delays, i.e that, during a fusion interval, they transmit their decisions to the fusion center with the same probability p, i.e., $p_1 = p_2 = p_3 = p$. Furthermore, and for the same reasons, it was assumed that the error probabilities in the different channels over which the sensors transmit their decision to the fusion, were the same. Under these assumptions, the probabilities of detection and false alarm at the k-th sensor are given by ([12] and [14])

$$P_{F_k} = [\lambda_k(1+\epsilon_k)]^{-1-\frac{1}{\epsilon_k}} \qquad \text{and} \qquad P_{D_k} = P_{F_k}^{\frac{1}{1+\epsilon_k}} \qquad (40)$$

where λ_k denotes the threshold of the k-th sensor and ϵ_k the signal-to-noise ratio at the k-th sensor. In the case of N=3, we have seven possible decision sets that may be received by the fusion center:

$$\begin{array}{lll} U_{2,3}^1 = \{u_1\} & U_3^{1,2} = \{u_1,u_2\} & U^{1,2,3} = \{u_1,u_2,u_3\} \\ U_{1,3}^2 = \{u_2\} & U_1^{2,3} = \{u_2,u_3\} & \\ U_{1,2}^3 = \{u_3\} & U_2^{1,3} = \{u_1,u_3\} & \end{array} \qquad (41)$$

with the following probabilities

$$P(U_{2,3}^1) = P(U_{1,3}^2) = P(U_{1,2}^3) = \frac{p(1-p)^2}{1-(1-p)^3} \qquad (42)$$

$$P(U_3^{1,2}) = P(U_1^{2,3}) = P(U_2^{1,3}) = \frac{p^2(1-p)}{1-(1-p)^3} \qquad (43)$$

$$P(U^{1,2,3}) = \frac{p^3}{1-(1-p)^3} \qquad (44)$$

The fusion center can choose different combining policies according to the number of peripheral decisions received. For example, if the set U contains only one decision, the fusion center must take this decision as true, since there is no other choice. If the set U contains two decisions, the fusion can choose either an AND or OR policy to combine the decisions. If U contains three decisions, the majority logic (ML) can also be used in addition to AND and OR. Simulation results from various fusion rules in slowly fading Rayleigh channels [12-13] are given next for different delays and channel error probabilities.

Figures 2 to 5 correspond to the ideal (errorless) channel cases. In Figures 2 and 3 the probability of detection at the fusion for both optimal and suboptimal cases with equal signal-to-noise ratios (SNR's) is plotted as a function of the sensor's SNR and the networking (delay) parameter p under three different combining policies. The probability of false alarm is fixed at 10^{-6}. It can be seen that the P_{D_0} is a monotonic increasing function of p for the OR-OR policy. However, this is not true if the decision rule AND-AND is chosen. It is worth noticing the "folding effect" in the case of the AND-AND fusion rule, meaning that higher delay may yield higher probability of detection for the same probability of false alarm. The "folding effect" is explained by noticing that the highest probability of detection at the fusion is achieved with only one sensor, if the fusion rule is AND. This in turns implies that, if the delays are high, i.e. the parameter p is small, the probability of having only one decision (or, a small subset of decisions) at the fusion at the time of fusing is higher, which renders the performance of the fusion closer to that of one sensor (optimal for AND).

A comparison of three decision rules (OR, AND and ML-OR) for p=0.9 is given in Figure 4. It is seen that the OR policy yields the best performance. Figure 5 gives the solutions for unequal SNR case with p=0.9 and P_{F_0} =10^{-6} under the OR policy. For all the cases discussed above, the sequential algorithm yields results which are very close to the optimal ones. Additional figures of the performance of different fusion rules as well as the optimal thresholds for the case of no channel errors can be found in [12] and [14].

Figures 6 to 8 correspond to the noisy channel cases. The probability of false alarm at the fusion is fixed at 10^{-4}. For each policy, two sets of curves are plotted; one set without delay, and another with delay. We assume that all three channels have the same probability of being in error, i.e. $P_{c_1} = P_{c_2} = P_{c_3} = P_c$. Computer simulation results have shown that if the OR policy is chosen, P_c must be kept at 10^{-5} level in order to achieve the desired $P_{F_0} = 10^{-4}$. However, for the AND policy, the same P_{F_0} can be achieved when P_c is at 10^{-3}.

CONCLUSIONS

In this paper we have shown that when delays due to networking or/and errors due to channel noise are introduced in the distributed detection problem of Figure 1, the optimal fusion rule is still the Neyman-Pearson test at all sensors and the fusion center. However, the thresholds at the sensors and at the fusion center are adjusted to account for the networking delays and the channel errors. Using the Lagrange multipliers formulation, the optimal set of thresholds is obtained in terms of a set of coupled nonlinear equations whose solution depends on the unknown fusion rule, thus it is impossible to solve for them except in trivial cases. The suboptimal algorithm that was developed in [12-13] and [14] has been modified to solve for the thresholds sequentially. Numerical comparison of the optimal and suboptimal solutions for both equal and unequal SNR cases has shown that the results given by the suboptimal algorithm are very close to the optimal ones.

It is also shown that, in the case of noisy channels, the decision made by each sensor depends on the reliability of the corresponding transmission channel. Moreover, the probability of false alarm at the fusion is restricted by the channel errors. For a given decision rule, the probability of any channel being in error must be kept at a certain level in order to achieve a desired probability of false alarm at the fusion. In the case of network delay, with or without channel errors, it is seen that the AND fusion rule yields poor performance. Furthermore, it suffers from a "folding effect", meaning that the performance of the fusion is not monotonic w.r.t. the values of the networking parameters (i.e., higher delay may yield higher probability of detection for the same probability of false alarm).

REFERENCES

[1] Conte, E., D'Addio, E., Farina, A. and Longo, M., "Multistatic Radar Detection: Synthesis and Comparison of Optimum and Suboptimum Receivers," IEEE Proc. F, Commun., Radar & Signal Process, 1983

[2] Tenney, R. R. and Sandell, N.R., Jr., "Detection with Distributed Sensors," IEEE Trans. on Aerospace and Electronic Systems, Vol. AES-17, July 1981, pp. 501-510.

[3] Sadjadi, F. A., "Hypothesis Testing in A Distributed Environment," IEEE Trans. on Aerospace and Electronic Systems, Vol. AES-22, March 1986, pp. 134-137.

[4] Teneketzis, D. and Varaiya, P. "The Decentralized Quickest Detection Problem," IEEE Trans. on Automatic Control, Vol. AC-29, No. 7, July 1984, pp. 641-644.

[5] Teneketzis, D., "The Decentralized Wald Problem," Proceedings of the IEEE 1982 International Large-Scale System Symposium, Virginia Beach, October 1982, pp. 423-430.

[6] Tsitsiklis, J. and Athans, M., "On the Complexity of Distributed Decision Problems," IEEE Trans. on Automatic Control, Vol. AC-30, No. 5, May 1985, pp. 440-446.

[7] Chair, Z. and Varshney, P. K., "Optimal Data Fusion in Multiple Sensor Detection Systems," IEEE Trans. on Aerospace and Electronic Systems, Vol. AES-22, No. 1, January 1986, pp. 98-101.

[8] Thomopoulos, S. C. A., Viswanathan, R. and Bougoulias, D. K., "Optimal Decision Fusion in Multiple Sensor Systems," IEEE Trans. on Aerospace and Electronic Systems, Vol.23, No. 5, Sept. 1987.

[9] Thomopoulos, S. C. A., Viswanathan, R. and Bougoulias, D. K., "Optimal Decision Fusion in Multiple Sensor Systems," Proceedings of the 24th Allerton Conference, October 1-3, 1986, Allerton House, Monticello, Illinois, pp. 984-993.

[10] Srinivasan, R., "Distributed Radar Detection Theory," IEE Proceedings, Vol. 133, Pt.F, No. 1, February 1986, pp. 55-60.

[11] Viswanathan, R. and Thomopoulos, S. C. A., "Distributed Data Fusion - A Singular Situation in Lagrange Multiplier Optimization," Technical Report, TR-SIU-DEE-87-4, Department of Electrical Engineering, Southern Illinois University, Carbondale, Illinois 62901, April 1987.

[12] Thomopoulos, S. C. A., Viswanathan, R. and Bougoulias, D.K., "Optimal Computable Distributed Decision Fusion," to be presented at the CDC'87, Los Angeles, Dec. 9-11, 1987.

[13] Thomopoulos, S. C. A., Viswanathan, R. and Bougoulias, D.K., "Optimal and Suboptimal Distributed Decision Fusion," submitted to IEEE Trans. on Aerospace and Electronic Systems.

[14] Thomopoulos, S. C. A. and Zhang, L., "Networking in Distributed Decision Fusion," SPIE Proceedings, Los Angeles, CA, Jan. 10-15, 1988, to appear.

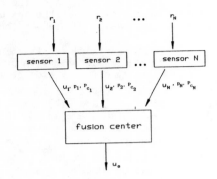

Fig 1   Distributed Fusion System

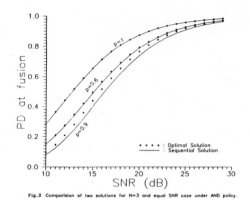

Fig.2 Comparision of two solutions for N=3 and equal SNR case under AND policy.

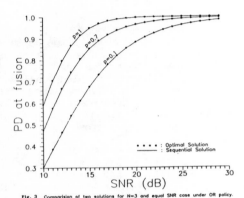

Fig. 3   Comparision of two solutions for N=3 and equal SNR case under OR policy.

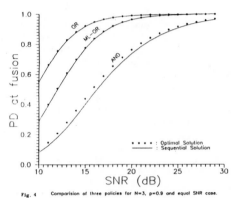

Fig. 4   Comparision of three policies for N=3, p=0.9 and equal SNR case.

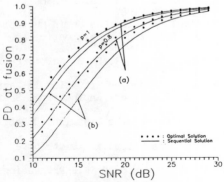

Fig. 5   Comparision of thresholds for N=3, p=0.9 and equal SNR case.

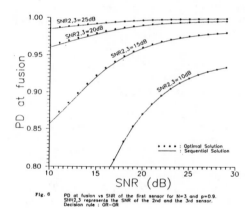

Fig. 6   PD at fusion vs SNR of the first sensor for N=3 and p=0.9.
SNR2,3 represents the SNR of the 2nd and the 3rd sensor.
Decision rule : OR-OR

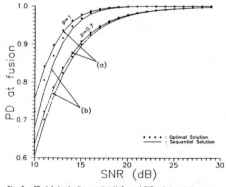

Fig. 7   PD at fusion for the case that N=3, equal SNR and channels have errors
Decision rule : AND-AND, PF=0.0001   (a) Pc=0.0001   (b) Pc=0.001

Fig. 8   PD at fusion for the case that N=3, equal SNR and channels have errors
Decision rule : OR-OR,   PF=0.0001   (a) Pc=0.00001   (b) Pc=0.00003

DISTRIBUTED FILTERING WITH RANDOM SAMPLING AND DELAY

Stelios C.A. Thomopoulos[1] and Lei Zhang
Department of Electrical Engineering
Southern Illinois University
Carbondale, IL 62901

A B S T R A C T

The problem of estimation and filtering in a distributed sensor environment is considered. The sensors obtain measurements about target trajectories at random times which transmit to the fusion center. The measurements arrive at the fusion with random delays which are due to queueing delays, and random delays in the transmission time as well as in the propagation time (sensor position may be unknown or changing with respect to the fusion). The fusion generates estimates of the target tracks using the received measurements. The measurements are received from the sensors at random times, they may have unknown time-origin and may arrive out of sequence. Optimal filters for the estimation problem of target tracks based on measurements of uncertain origin received by the fusion at random times and out of sequence have been derived for the cases of random sampling, random delay, and both random sampling and random delay. It is shown that the optimal filters constitute an extension to the Kalman Filter to account for the uncertainty involved with the data time-origin.

INTRODUCTION

The problem of target track estimation in the presence of uncertainty has received a lot of attention recently and constitutes a realistic framework for distributed surveillance systems. In distributed sensor systems, geographically dispersed sensors monitor targets by sampling their tracks. Subsequently, either they estimate the target position (and/or velocity) and transmit their estimates to a fusion center, or, they transmit their measurements directly to the fusion center. The fusion center is responsible for combining the estimates or the measurements from the sensors into the final decision about the target position and/or velocity.

In a distributed sensor environment the quality of the estimates depends on the number of sensors, the relative sampling times of the sensors, the uncertainty about the origin of the samples and the delays (random or nonrandom) associated with the transmission of the data from the sensors to the fusion.

In the literature [7], [8], it is often assumed that the sensors transmit their estimates to the fusion instead of the actual measurements. However, in order for the fusion to combine the estimates from the different sensors, the error covariances as computed by the different sensors are required and must be transmitted to the fusion which defeats the advantage of parallel processing of the measurements from the different sensors by increasing the speed at which data can be transmitted from the sensors to the fusion center. In the cases to be discussed in this paper the error covariance of the filters depend on the data and cannot be precomputed. Hence, there is no advantage in preprocessing the measurements at the sensors level and combining the estimates at the fusion level because the transmission of the error covariance is required. Thus, in the cases that are discussed next, we assume that the sensors transmit their actual measurements and not their estimates to the fusion.

Problems related to estimation with data of uncertain origin have been studied by Nahi [1], Jaffer and Gupta [2], [3],. The problem of probabilistic data association with uncertainty in the origin of the data was studied by Bar-Shalom and Marcus [4]. However, both their scope and their filter differ from ours in that their approach is Bayesian and requires computation of the a-posteriori probability distribution of the origin uncertainty. Rhodes and Snyder [5] have considered a problem of combined estimation and control with measurements occurring at random times.

The paper is structured as follows. For a fixed number of sensors, we compare first the performance of a "parallel" scheme, where all the sensors are synchronized and sample the target tracks at the same time instances at the beginning of equal sampling intervals of normalized length one, with a "serial" scheme, where the sensors sample at equally spaced subdivisions of the same normalized sampling intervals. The comparison of the parallel with the serial scheme is done for reference reasons so that the performance of the filters with uncertainty in the sampling time and delays is judged against. In the sequel we introduce uncertainty about the sampling origin and the delays, for which cases we derive optimal and suboptimal filters, when the optimal ones are not realizable.

---

1 This research is sponsored by SDIO/IST and managed by the Office of Naval Research Under Contract N00014-86-k-0515.

The problem of estimation and filtering in a distributed sensor environment is considered. The sensors obtain measurements about target trajectories at random times which they transmit to the fusion center. The measurements arrive at the fusion with random delays which are due to random delays in the transmission time, as well as in the propagation time (sensor position may be unknown or changing with respect to the fusion). The fusion generates estimates of the target tracks using the received measurements. The measurements are received from the sensors at random times, may have unknown time-origin and may arrive out of sequence. It is shown that the optimal filter for the estimation problem of target tracks based on measurements of uncertain origin received by the fusion at random times and out of sequence, is an extension to the Kalman Filter to account for the uncertainty involved with the data time-origin.

Three different operating scenarios are considered. The first involves sampling at random times and transmission over a channel with fixed delay. The second involves sampling at fixed time instances but transmission over a channel with random delay. The third scenario involves random sampling and random transmission delay. For all three scenarios, the minimum mean-squared error filters are derived assuming that the process and observation noises are both Gaussian. For the case of random sampling and random delay, the optimal filter is non-realizable. Hence, a suboptimal filter with fixed memory and limited complexity is derived. The performance of all the filters is analyzed and tested numerically.

First we describe the model that we use for the derivation of the optimal filter. For this model, the filter equations are derived assuming that both the process and the observation equations are linear and time-invariant. The results that are obtained generalize easily to time-varying linear systems, provided that the proper transition matrices and filter equations are used.

I. The Model

We assume that the processes whose states are being estimated is described by a system of continuous-time, first order, linear stochastic differential equations

$$\dot{x}(t) = F x(t) + D \tilde{w}(t) \tag{1}$$

where

x is the state vector of dimension Mx1,
w is the process noise which is assumed to be white, gaussian with zero-mean and covariance

$$E\{\tilde{w}(t) \ \tilde{w}(\tau)\} = q(t) \ \delta(t-\tau) \tag{2}$$

F, D are known constant matrices of dimension M x M and M x 1, respectively.
Suppose that there are N sensors in the system. The measurement model at the ith sensor is assumed to be given by

$$z_i(t) = H_i x(t) + \tilde{v}_i(t) \qquad\qquad i = 1, 2, \ldots, N \tag{3}$$

where

z_i is a vector of dimension L_i,
v_i is the observation noise which is assumed to be white, gaussian with zero mean and covariance

$$E\{\tilde{v}_i(t) \ \tilde{v}_i^T(\tau)\} = V_i(t) \ \delta(t-\tau) \tag{4}$$

and

H_i is a known constant matrix of dimension L_i x M.
The noises $\tilde{w}(t)$ and $\tilde{v}_i(t)$'s are assumed to be mutually uncorrelated, thus mutually independent since gaussian.

Remarks: 1. The filters that are derived in the sequel generalize easily to time varying matrices F, D, and H_i provided that the appropriate changes are made in the state transition matrices and the error covariance matrices [6].

2. If the assumption of gaussianness is removed from the model assumptions, the derived filters are optimal among all linear filter.

Since the data sampling and processing are done discretely in time, we first discretize the process and observation equations before we derive the optimal discrete-time filters. In discrete time, the dynamic equation is written as

$$x_{t_{k+1}} = \Phi(t_k, t_{k+1}) x_{t_k} + w_{t_k} \tag{5}$$

and the measurement equations are

$$z_{it_k} = H_i x_{t_k} + v_{it_k} \qquad\qquad i = 1, 2, \ldots, N \tag{6}$$

where

$$\Phi(t_k, t_{k+1}) = e^{F(t_{k+1} - t_k)} \tag{7}$$

$$w_{t_k} = \int_{t_k}^{t_{k+1}} e^{F(t_{k+1} - \tau)} D \tilde{w}(\tau)d\tau \tag{8}$$

w_{t_k} and v_{it_k}'s are the discrete-time process and measurement noises, they are zero-mean white random sequences with covariance

$$E\{ w_{t_k} w_{t_k}^T \} = Q(t_k) \ \delta_{kj} \tag{9}$$

$$E\{ v_{it_k} v_{it_j}^T \} = R_i(t_k) \ \delta_{kj} \qquad\qquad i = 1, 2, \ldots, N \tag{10}$$

where δ_{kj} is Kronecker's delta function and

$$Q(t_k) = \int_{t_k}^{t_{k+1}} \Phi(t_{k+1}, \tau)\, D\, q(\tau)\, D^T\, \Phi^T(t_{k+1},\tau)d\tau \qquad (11)$$

$$R_i(t_k) = \frac{V_i(t_k)}{t_{k1} - t_k} \qquad\qquad i = 1, 2, \ldots, N \qquad (12)$$

In the derivation of the different filters we assume that the discrete model, Eq.'s (5) and (6), is used.

Parallel vs. Serial Sampling

When all the sensors are synchronized and they all sample at the beginning of fixed-length time intervals, different implementations of the optimal filter at the fusion are possible. The optimal filter at the fusion can be implemented either by processing all the data in a batch [6], or by processing the data sequentially, or by processing the intermediate estimates from the sensors sequentially [7], [8]. However, when the estimates from the sensors are combined to yield the optimal filter, the error covariance matrix of each sensor is required. Furthermore, at the fusion level the combined error covariance matrix need to be computed once all the covariance matrices are received. Hence, there is no obvious advantage in processing speed from computing the local estimates at the sensor level first. Furthermore, transmission of the error covariance matrices reduces the transmission rate if the matrices are transmitted by the sensors, or reduces the computational efficiency if they are recomputed at the fusion for each sensors. For all these reasons, it is no obvious weather there is indeed any advantage from parallel processing of the measurements at the sensors and combination of the estimates at the fusion. Thus, in the derivation of all the filters we assume that the sensors transmit their measurements to the fusion directly without any preprocessing.

In the serial case, we assume that the sampling interval used in the parallel filtering, is subdivided into a number of equal-sized subintervals equal to the number of the sensors and that each sensor samples at the beginning of one and only one of these subintervals, Fig. 1. The sampling subinterval for a given sensor remains fixed throughout the entire filtering period. As it will be seen from equations (11) and (12), subdivision of the sampling interval into equal-length intervals, allows a direct comparison of the performance of the parallel and serial schemes at the end of each sampling interval (see Simulation Results Section), since the process and measurement covariance matrices remain the same in both cases.

The implementation of the serial filter is done according to the standard Kalman Filter [6], except that when a new measurement arrives the previous estimate is extrapolated first at the time instant the measurement was obtained (we assume zero transmission delay in this case). After the measurement is obtained, the estimate is updated according to the update equation of the standard Kalman Filter. For the case of the serial filter with zero propagation delay, an alternative implementation using the Kalman Filter estimates of each sensor at the fusion is feasible and is given [10]. However, its computational complexity renders the implementation of the estimate-based filter unattractive.

Comparison of the performance of the parallel and serial schemes is done in the Simulation Results Section. In order to obtain a meaningful comparison, the estimates and the error covariance matrices are extrapolated at the beginning of the sampling intervals, provided that the same number of samples from each sensor are used.

In a realistic distributed sensing environment, perfect synchronization of the sampling times is practically impossible. Furthermore, there is always random delay associated with the transmission of the data (measurements or estimates) from the sensors to the fusion. Hence, the assumptions of perfect sampling times and zero transmission delays must be relaxed, in order to obtain more realistic filters and more accurate assessment of their performance under more realistic assumptions about the operational conditions in distributed sensor-fusion systems. We derive the optimal filters under the realistic assumptions of random sampling and random delay in the transmission. For the sake of clarity, we introduce randomness one step at a time. That is, we first relax the assumption of perfect sampling times and introduce uncertainty about the sampling time while we keep the propagation delays fixed. Next we assume that the sampling is perfect, but there is random transmission delay. Finally, we relax both the assumptions and we consider the general case with random sampling and random transmission delay. In all cases we assume that the uncertainty in either the sampling time or the delay, enter uniformly about some nominal value used in the absence of uncertainty.

II. Random Sampling and Fixed Delay

The actual time-axis is divided in time intervals of fixed length T. Let T be the sampling interval that the sensors use to obtain data (measurements from the targets). Each sensor is assumed to obtain measurements at random only once in every time interval [kT, (k+1)T), k = 1, 2, 3, ..., according to some known distribution and transmit the data to the fusion through a channel with fixed delay Δ, which is the same for all the sensors. The fusion receives the data which are sampled at times $t_1-\Delta$, $t_2-\Delta$, ..., $t_k-\Delta$, ... at times $t_1, t_2, \ldots, t_k, \ldots$, respectively, Fig. 2. The optimal filter is linear and has the following form

$$\hat{x}_{t_k}(+) = K' \hat{x}_{t_k}(-) + K\, z_{t_k-\Delta} \qquad (13)$$

where

$\hat{x}_{t_k}(+)$ is the estimate of state x_{t_k} based on the data received up to time t_k.

$\hat{x}_{t_k}(-)$ is the prediction of state x_{t_k} based on the data received up to time t_{k-1}.

$z_{t_k - \Delta}$ is the measurement received at time t_k, which could have come from any of the N sensors, so it will be denoted by H_c, c standing for current. Similarly, the measurement noise covariance matrix of the current data will be denoted by R_c. We determine the matrices K and K' next.

Since

$$x_{t_k} = \Phi(t_k-\Delta, t_k) \, x_{t_k-\Delta} + w_{t_k-\Delta} \tag{14}$$

$$z_{t_k-\Delta} = H_c \, x_{t_k-\Delta} + v_{t_k-\Delta} \tag{15}$$

we have

$$x_{t_k-\Delta} = \Phi^{-1}(t_k-\Delta, t_k) [x_{t_k} - w_{t_k-\Delta}] \tag{16}$$

So,

$$\hat{x}_{t_k}(+) = K' \hat{x}_{t_k}(-) + K[H_c \, x_{t_k-\Delta} + v_{t_k-\Delta}] \tag{17}$$

$$= K' \hat{x}_{t_k}(-) + K H_c \, \Phi^{-1}(t_k-\Delta, t_k) \, [x_{t_k} - w_{t_k-\Delta}] + K v_{t_k-\Delta}$$

Define

$$\tilde{x}_{t_k}(+) = \hat{x}_{t_k}(+) - x_{t_k} \tag{18}$$

$$\tilde{x}_{t_k}(-) = \hat{x}_{t_k}(-) - x_{t_k} \tag{19}$$

Then

$$\tilde{x}_{t_k}(+) = K' \tilde{x}_{t_k}(-) + [K' - I + K H_c \, \Phi^{-1}(t_k-\Delta,t_k) \, x_{t_k}] - K H_c \, \Phi^{-1}(t_k-\Delta,t_k) w_{t_k-\Delta} + K v_{t_k-\Delta} \tag{20}$$

The estimate is unbiased if and only if

$$K' - I + K H_c \, \Phi^{-1}(t_k-\Delta,t_k) = 0 \tag{21}$$

Thus, we obtain

$$K' = I - K H_c \, \Phi^{-1}(t_k-\Delta,t_k) \tag{22}$$

and

$$\hat{x}_{t_k}(+) = \hat{x}_{t_k}(-) + K [z_{t_k-\Delta} - H_c \, \Phi^{-1}(t_k-\Delta,t_k) \, \hat{x}_{t_k}(-)] \tag{23}$$

To determine the filter gain K, define the error covariance matrix

$$P_{t_k}(+) = E\{\tilde{x}_{t_k}(+) \, \tilde{x}_{t_k}^T(+)\} \tag{24}$$

$$P_{t_k}(-) = E\{\tilde{x}_{t_k}(-) \, \tilde{x}_{t_k}^T(-)\} \tag{25}$$

Defining

$$\Phi_\Delta =: \Phi(t_k, t_k-\Delta)$$

in order to keep the expressions more compact, $P_{t_k}(+)$ can be expressed as

$$P_{t_k}(+) = (I - KH_c\Phi_\Delta^{-1}) \, P_{t_k}(-) \, (I - KH_c\Phi_\Delta^{-1}) + Q_{t_k-\Delta} \, (KH_c\Phi_\Delta^{-1})^T$$

$$+ KH_c\Phi_\Delta^{-1} Q_{t_k-\Delta} + KR_{t_k-\Delta} K^T - KH_c\Phi_\Delta^{-1} \, Q_{t_k-\Delta} \, (KH_c\Phi_\Delta^{-1})^T \tag{26}$$

Note that

$$E\{\tilde{x}_{t_k}(-) \, w_{t_k-\Delta}^T\} = E\{(\hat{x}_{t_k}(-) - x_{t_k}) \, w_{t_k-\Delta}^T\} = - E\{x_{t_k} w_{t_k-\Delta}^T\}$$

$$= - E\{[\Phi(t_k-\Delta, t_k) \, x_{t_k-\Delta} + w_{t_k-\Delta}] w_{t_k-\Delta}^T\}$$

$$= - E\{w_{t_k-\Delta} w_{t_k-\Delta}^T\} = - Q_{t_k-\Delta} \tag{27}$$

$$= E\{w_{t_k-\Delta} \, \tilde{x}_{t_k}(-)^T\}$$

Minimizing $P_{t_k}(+)$ with respect to K, and defining

$$K_{t_k} =: K ,$$

to indicate the dependence of the filter gain on the arrival time, we obtain

$$K_{t_k} = [P_{t_k}(-) - Q_{t_k-\Delta}](H_c\Phi_\Delta^{-1})^T \{H_c\Phi_\Delta^{-1}[P_{t_k}(-) - Q_{t_k-\Delta}](H_c\Phi_\Delta^{-1})^T + R_{t_k-\Delta}\}^{-1} \tag{28}$$

$$P_{t_k}(+) = [I - K_{t_k} H_c\Phi_\Delta^{-1}] [P_{t_k}(-) - Q_{t_k-\Delta}] + Q_{t_k-\Delta} \tag{29}$$

or, equivalently,

$$[P_{t_k}(+) - Q_{t_k-\Delta}]^{-1} = [P_{t_k}(-) - Q_{t_k-\Delta}]^{-1} + (H_c\Phi_\Delta^{-1})^T R_{t_k-\Delta}^{-1}(H_c\Phi_\Delta^{-1}) \tag{30}$$

$$K_{t_k} = [P_{t_k}(+) - Q_{t_k-\Delta}] [H_c\Phi_\Delta^{-1}(t_k-\Delta, t_k)]^T R_{t_k-\Delta}^{-1} \tag{31}$$

The state prediction $\hat{x}_{t_k}(-)$ and its error covariance $P_{t_k}(-)$ can be obtained by

$$\hat{x}_{t_k}(-) = \Phi(t_{k-1}, t_k) \, \hat{x}_{t_{k-1}}(+) \tag{32}$$

$$P_{t_k}(-) = \Phi(t_{k-1}, t_k) \, P_{t_{k-1}}(+) \, \Phi^T(t_{k-1}, t_k) + Q_{t_{k-1}} \tag{33}$$

Equations (23), (28) and (29), or (30) and (31), and the prediction equations (32) and (33), constitute the equations of the optimal filter for the case of random sampling and fixed delay. Notice that the in the estimate equation (23), the correction term (second term in the bracket, in the r.h.s. of the equation) is a white, gaussian term of time reference $t_k-\Delta$ lagging by Δ w.r.t. the present time t_k.

III. Fixed Sampling and Random Delay

Suppose that each of the N sensors samples once every T time units, and that the sensors sample sequentially in a fixed sequence, i.e. sensor i samples at iT/N, T+iT/N, 2T+iT/N, ..., etc. The data are transmitted to the fusion over channels with random delays. Assuming that the delays Δ_k, k = 1, 2, ..., N, are uniformly distributed in $[0, 2T/N]$, the data received by the fusion center may be out of sequence in the sense that earlier samples may be received after subsequent ones, Fig. 3. From the arrival time and the fixed sampling times, the delay encountered by each data can be computed exactly. Hence, from the filter point-of-view, the delay can be computed deterministically in this case.

If the data received at time t_k was sampled after the data received at t_{k-1}, the noises in the measurements from the corresponding sensors are uncorelated and the optimum filter has the same form as the one in the random sampling and fixed delay case, Fig. 1. If, however, the data received at time t_k was sampled before the data received at t_{k-1}, Fig. 3, the noises in the measurements from the corresponding sensors are correlated. Then

$$E\{x_{t_k}(-) \ w_{t_k-\Delta_k}^T\} = E\{[x_{t_k}(-) - x_{t_k}] \ w_{t_k-\Delta_k}^T\} \tag{34}$$
$$= E\{\hat{x}_{t_k}(-) \ w_{t_k-\Delta_k}\} \leq Q_{t_k-\Delta_k}$$
$$= E\{\Phi(t_{k-1}, t_k) \ \hat{x}_{t_{k-1}}(+) \ w_{t_k-\Delta_k}\} - Q_{t_k-\Delta_k}$$
$$= \Phi(t_{k-1}, t_k) \ E\{x_{t_{k-1}}(+) \ w_{t_k-\Delta_k}\} - Q_{t_k-\Delta}$$

Since
$$\hat{x}_{t_{k-1}}(+) = \hat{x}_{t_{k-1}}(-) + K_{t_{k-1}} [z_{t_{k-1}-\Delta_{k-1}} - H_p\Phi^{-1}(t_{k-1}-\Delta_{k-1}, t_{k-1})\hat{x}_{t_k}(-)] \tag{35}$$
and
$$E\{\hat{x}_{t_{k-1}}(-) \ w_{t_k-\Delta_k}^T\} = 0 \tag{36}$$
we have
$$E\{\hat{x}_{t_{k-1}}(+) \ w_{t_k-\Delta_k}^T\} = K_{t_{k-1}} \ E\{z_{t_{k-1}-\Delta_{k-1}} \ w_{t_k-\Delta_k}^T\}$$
$$= K_{t_{k-1}} \ E\{[H_p x_{t_{k-1}-\Delta_{k-1}} + w_{t_{k-1}-\Delta_{k-1}}] \ w_{t_k-\Delta_k}^T\}$$
$$= K_{t_{k-1}} H_p E\{x_{t_{k-1}-\Delta_{k-1}} \ w_{t_k-\Delta_k}\}$$
$$= K_{t_{k-1}} H_p E\{[\Phi(t_{k-\Delta_k}, t_{k-1}-\Delta_{k-1}) \ x_{t_k-\Delta_k} + w_{t_k-\Delta_k}] \ w_{t_k-\Delta_k}^T\}$$
$$= K_{t_{k-1}} H_p \ Q_{t_k-\Delta_k} \tag{37}$$

So,
$$E\{\tilde{x}_{t_k}(-) \ w_{t_k-\Delta_k}\} = - [I - \Phi(t_{k-1}, t_k) \ K_{t_{k-1}} H_p] \ Q_{t_k-\Delta_k}$$
$$= - Q'_{t_k-\Delta_k} \tag{38}$$
where
$$Q'_{t_k-\Delta_k} = [I - \Phi(t_{k-1}, t_k) K_{t_{k-1}} H_p] Q_{t_k-\Delta_k} \tag{39}$$

Replacing $Q_{t_k-\Delta_k}$ by $Q'_{t_k-\Delta_k}$ in Eqs. (28) and (29) and defining for notational compactness
$$\Phi_{\Delta_k} =: \Phi(t_k-\Delta_k, t_k) ,$$
we obtain the optimal filter
$$x_{t_k}(+) = x_{t_k}(-) + K_{t_k} [z_{t_k-\Delta_k} - H_c\Phi^{-1}(t_k-\Delta_k, t_k) \ \hat{x}_{t_k}(-)] \tag{40}$$
and the gain and error covariance update equations
$$K_{t_k} = [P_{t_k}(-) - Q'_{t_k-\Delta}] (H_c\Phi^{-1})^T \{H_c\Phi^{-1}[P_{t_k}(-) - Q'_{t_k-\Delta}] (H_c\Phi^{-1})^T + R_{t_k-\Delta}\}^{-1} \tag{41}$$
$$P_{t_k}(+) = [I - K_{t_k}H_c\Phi^{-1}] [P_{t_k}(-) - Q'_{t_k-\Delta}] + Q'_{t_k-\Delta} \tag{42}$$

$$\text{for } t_k-\Delta_k < t_{k-1}-\Delta_{k-1}$$

It is worth noticing the switching phenomenon between the two filters in the case of fixed sampling and random delay. When the data is received in sequence, i.e. first sampled first received, the filter equations are equations (23), (28) and (29), or (30) and (31), and the prediction equations (32) and (33). However, if the data arrive out of sequence, then the filter equations are (40), (41) and (42), along with the propagation equations (32) and (33).

IV. Random Sampling and Random Delay

In the random sampling and random delay case, the optimal filter is non-realizable. Since both the sampling time and the delay are random and unknown, neither one can be calculated precisely from their sum, i.e. the arrival time at the fusion. Hence, in order to determine weather the current measurement is correlated or uncorelated with respect to the previous one, the filter must be conditioned on either the sampling time or the delay, from which the other unknown random variable (delay or sampling time respectively) can be obtained since their sum (the arrival time) is known. The conditional optimal filter with fixed sampling (delay) and random delay (sampling) can be obtained provided that it is possible to determine weather the current measurement is correlated or uncorelated w.r.t. the previous one. However, in order to determine the correlation between the two consecutive measurements, the sampling time of the previous measurement need to be known, which is not. Hence, one should condition the previous estimate on the sampling time of the previous measurement and obtain the update estimate conditioned on both current and previous sampling times, form which the update estimate may be obtained by averaging over the joint density of both sampling times. However, since both sampling time random variables are continuous, a non-countable infinite number of conditional filters would be required making the optimal filter non-realizable. In order to obtain a realizable filter, the sampling time of every measurement is set equal to its conditional a-posteriori mean.

This suboptimal filter can be realized using the fixed sampling and random delay filter as it is described next.

Let the sampling time t_{sk}, $k=1,2,\ldots$, be uniformly distributed in the interval $[kT/N, (k+1)T/N]$, i.e., the probability density function be

$$p_s(t_{sk}) = \begin{cases} \frac{1}{T/N} & \text{for } kT/N \le t_{sk} \le (k+1)T/N \\ 0 & \text{otherwise} \end{cases} \tag{43}$$

and let the corresponding delay Δ_k, $k=1,2,\ldots$, be uniformly distributed in the interval $[t_{sk}, t_{sk}+ T/N]$, i.e.,

$$p_\Delta(\Delta_k) = \begin{cases} \frac{1}{T/N} & \text{for } t_{sk} \le \Delta_k \le t_{sk}+ T/N \\ 0 & \text{otherwise} \end{cases} \tag{44}$$

Then the arrival time t_k, $k=1,2,\ldots$, at which the fusion center receives data is given by

$$t_k = t_{sk} + \Delta_k \tag{45}$$

The joint probability density function of the sampling time and the arrival time is

$$p(t_{sk},t_k) = p_s(t_{sk})p_\Delta(t_k-t_{sk}) = \begin{cases} \frac{N^2}{T^2} & \text{for } kT/N \le t_{sk}\le (k+1)T/N \text{ and } \\ & t_{sk}\le t_k \le t_{sk}+ T/N \\ 0 & \text{otherwise} \end{cases} \tag{46}$$

Integrate Eq. (46) over t_{sk}, we have

$$p_a(t_k) = \begin{cases} N^2[t_k- kT/N]/T^2 & \text{for } kT/N \le t_k\le (k+1)T/N \\ N^2[(k+2)T/N - t_k]/T^2 & \text{for } (k+1)T/N \le t_k \le (k+2)T/N \\ 0 & \text{otherwise} \end{cases} \tag{47}$$

Thus, the conditional density of the sampling time given the arrival time is

$$p_{s/a}(t_{sk}| t_k) = \frac{p(t_{sk},t_k)}{p_a(t_k)} = \begin{cases} 1/[t_k- kT/N] & \text{for } kT/N \le t_k\le (k+1)T/N, kT/N \le t_{sk}\le t_k \\ 1/[(k+2)T/N - t_k] & \text{for } (k+1)T/N \le t_k \le (k+2)T/N, \\ & t_k- T/N \le t_{sk}\le (k+1)T/N \end{cases} \tag{48}$$

Let $\bar{t}_{sk}$ denote the conditional mean of t_{sk} given t_k, then

$$\bar{t}_{sk} = \int_{kT/N}^{(k+1)T/N} t_{sk}\, p_{s/a}(t_{sk}| t_k)\, dt_{sk} = [t_k+ kT/N]/2 \tag{49}$$

where $kT/N \le t_k \le (k+2)T/N$

The equations for the suboptimal filter can be obtained by replacing the actual sampling time $t_k- \Delta$ with the conditional mean $\bar{t}_{sk}$ in Eqs. (28), (29) and (40) - (42), and are as follows.

Suboptimal Filter Equation:

$$\hat{x}_{t_k}(+) = \hat{x}_{t_k}(-) + K_{t_k} [z_{t_k} - H_c\Phi^{-1}(\bar{t}_{sk}, t_k) \hat{x}_{t_k}(-)] \tag{50}$$

Filter Gain and Error Covariance Update Equations:

(A) For $\bar{t}_{sk} > \bar{t}_{sk}^{k-1}$

$$K_{t_k} = [P_{t_k}(-) - Q_{\bar{t}_{sk}}^{k-1}] (H_c\Phi^{-1})^T \{H_c\Phi^{-1}[P_{t_k}(-) - Q_{\bar{t}_{sk}}] (H_c\Phi^{-1})^T+ R_{\bar{t}_{sk}} \}^{-1} \tag{51}$$

$$P_{t_k}(+) = [I - K_{t_k} H_c\Phi^{-1}] [P_{t_k}(-) - Q_{\bar{t}_{sk}}] + Q_{\bar{t}_{sk}} \tag{52}$$

(B) For $\bar{t}_{sk} < \bar{t}_{sk}^{k-1}$

$$K_{t_k} = [P_{t_k}(-) - Q_{\bar{t}_{sk}}^{k-1}] (H_c\Phi^{-1})^T \{H_c\Phi^{-1}[P_{t_k}(-) - Q'_{\bar{t}_{sk}}] (H_c\Phi^{-1})^T+ R_{\bar{t}_{sk}} \}^{-1} \tag{53}$$

$$P_{t_k}(+) = [I - K_{t_k} H_c\Phi^{-1}] [P_{t_k}(-) - Q'_{\bar{t}_{sk}}] + Q'_{\bar{t}_{sk}} \tag{54}$$

where

$$Q'_{\bar{t}_{sk}} = [I - \Phi(t_{k-1},t_k) K_{t_{k-1}} H_p] Q_{\bar{t}_{sk}} \tag{55}$$

State and Error Covariance Extrapolation Equations:

$$\hat{x}_{t_k}(-) = \Phi(t_{k-1},t_k) \hat{x}_{t_{k-1}}(+) \tag{56}$$

$$P_{t_k}(-) = \Phi(t_{k-1}, t_k) P_{t_{k-1}}(+) \Phi^T(t_{k-1}, t_k) + Q_{t_{k-1}} \tag{57}$$

SIMULATION RESULTS

The Kalman Filter with serial sampling was tested using the discrete model for a moving target with constant acceleration and the results were compared with the Kalman Filter with parallel sampling. The same model was used to test the random sampling and fixed delay filter, and the fixed sampling and random delay filter. At the time the present was written, no numerical results were available for the random sampling and random delay filter.

The model of moving target with constant acceleration (more precisely, random acceleration) that was used assumes that the continuous-time dynamics of the target can be described by the state equation [6]

$$\begin{vmatrix} \dot{x}_1(t) \\ \dot{x}_2(t) \end{vmatrix} = \begin{vmatrix} 0 & 1 \\ 0 & 0 \end{vmatrix} \begin{vmatrix} x_1(t) \\ x_2(t) \end{vmatrix} + \begin{vmatrix} 0 \\ 1 \end{vmatrix} w(t) \tag{58}$$

along with the observation equation

$$y(t) = | \ 1 \quad 0 \ | \ x^i(t) + v(t) \tag{59}$$

where $x(t)$ is the 2x1 state vector, $q(t)$ and $r(t)$ are uncorelated, zero mean, white gaussian, noises with covariances $q(t)\delta(t-\tau)$ and $r(t)\delta(t-\tau)$ respectively, with $q(t) \geq 0$ and $r(t) > 0$ for all t. The initial conditions on $x(t)$ are taken to be gaussian with mean $x(0)$ and positive definite covariance matrix $P(0)$.

Assuming that the acceleration remains constant during the k-th sampling interval and discretizing equations (58) and (59) every Δ time units, the discrete equivalent model is obtained [8]:

$$\begin{vmatrix} \dot{x}_1(k\Delta) \\ \dot{x}_2(k\Delta) \end{vmatrix} = \begin{vmatrix} 1 & T \\ 0 & 1 \end{vmatrix} \begin{vmatrix} x_1(k\Delta) \\ x_2(k\Delta) \end{vmatrix} + \omega(k\Delta) \tag{60}$$

along with the observation equation

$$y(k\Delta) = | \ 1 \quad 0 \ | \ x^i(k\Delta) + v(k\Delta) \tag{61}$$

where the noise covariance for $\omega(k\Delta)$ is [6], [8],

$$Q = \begin{vmatrix} T^4/4 & T^3/2 \\ T^3/2 & T^2 \end{vmatrix} q \ , \text{ assuming that } q(t) = \text{constant} = q, \tag{62}$$

and $r(k\Delta)/\Delta$ for $v(k\Delta)$.

Comparison of the performance of the serial Kalman Filter with the parallel one, indicates that the serial yields superior results. In order to obtain a meaningful comparison, the estimates and their error covariances must be compared at time-points corresponding to the same number of samples. From Figures 4, 5, 6 and 7 it is seen that error covariance of the the serial filter is substantially lower than the error covariance of the parallel filter and the estimates closer to the actual states.

Simulation results from the runs of the random sampling, fixed delay filter and the fixed sampling, random delay filters are shown in Figures 8, 9, 10, 11, 12 and 13. From these figures it is seen that the randomness due to either sampling or delay degrades the performance of the filters. Furthermore, although the dynamics of the state equations are stationary, the covariance matrix does not reach a steady-state due to the time variations in the observation model that are introduced by the uncertainty in the the sampling time or/and the transmission delay. Moreover, from Figs. 12 and 13 it is seen that random delays yield larger error covariance, i.e. they have a more negative effect on the quality of the filter than random sampling with fixed delays. Similar results are expected from the random sampling, random delay filters. For a more meaningful comparison of the numerical results and assessment of the degradation of the performance of the filters due to random sampling and random delays, the numerical numerical results must be extrapolated at time-points corresponding to the same number of collected samples (measurements) from the sensors. This constitutes an ongoing project and the results are to be announced in [9].

CONCLUSION

The problem of distributed filtering with random sampling and random transmission delays in the data from the sensors to the fusion center has been considered. Optimal filters for three different cases have been derived and tested in limited numerical cases.

REFERENCES

[1] Nahi, N. E., "Optimal recursive estimation with uncertain observation," IEEE Trans. Info. Theory, Vol. IT-15, pp. 457-462, July 1969.

[2] Jaffer, A. G. and Gupta, S. C., "Recursive Bayesian Estimation with Uncertain Observation," IEEE Trans. Inf. Theory, Vol. IT-17, pp. 614-616, Oct. 1971.

[3] Jaffer, A. G. and Gupta, S. C., "Optimal Sequential Estimation of Discrete Processes with Markov Interrupted Observation," IEEE Trans. Automat. Contr., vol. Ac-16, pp. 471-475, Oct. 1971.

[4] Bar-Shalom, Y. and Marcus, G. D.,Tracking with Measurements of Uncertain Origin and Random Arrival Times," IEEE Trans. Automat. Contr., Vol. AC-25, No. 4, pp. 802-807, Aug. 1980.

[5] Rhodes, I. B. and Snyder, D. L., "Estimation and Control Performance for Space-Time Point-Process Observations," IEEE Trans. Autom. Contr., Vol. AC-22, pp. 338-345, June 1977.

[6] Applied Optimal Estimation, Ed. A. Gelb, The M.I.T. Press, Sixth Edition, 1980, Cambridge, Mass.

[7] Bar-Shalom, Y., "Tracking Methods in a Multitarget Environment," IEEE Trans. Autom. Contr., Vol. AC-23, pp.618-626, Aug. 1978.

[8] Multiterget/Multisensor Tracking, Lecture Notes, UCLA Extension Lecture Series, Coordinator: Y. Bar-Shalom, 1986.

[9] Thomopoulos, S. C. A. and Zhang, L., "Distributed Estimation in the Presence of Uncertainty and Delays," 8th Digital Avionics Systems Conference, October 17, 1988, San Jose, Ca.

[10] Thomopoulos, S. C. A., "Distributed Estimation and Filtering," Technical Report TR-SIU-11-88, Southern Illinois University, Carbondale, IL, February 1988.

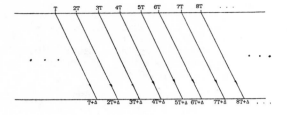

Figure 1   Serial Sampling Signal Transmissions

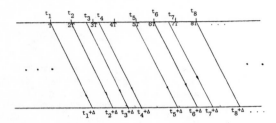

Figure 2   Random Sampling, Fixed Delay Signal Transmissions

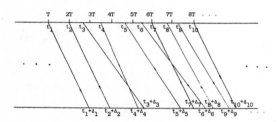

Figure 3   Fixed Sampling, Random Delay Signal Transmissions

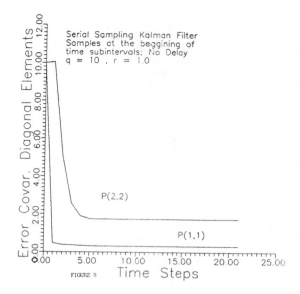

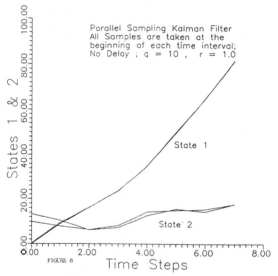

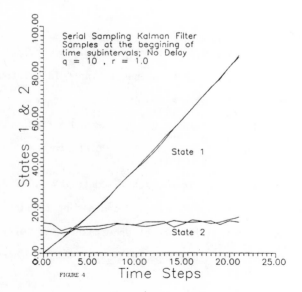

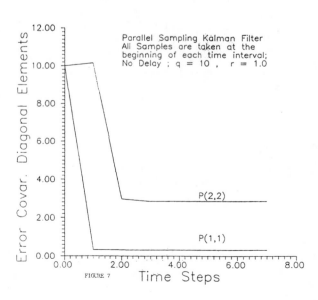

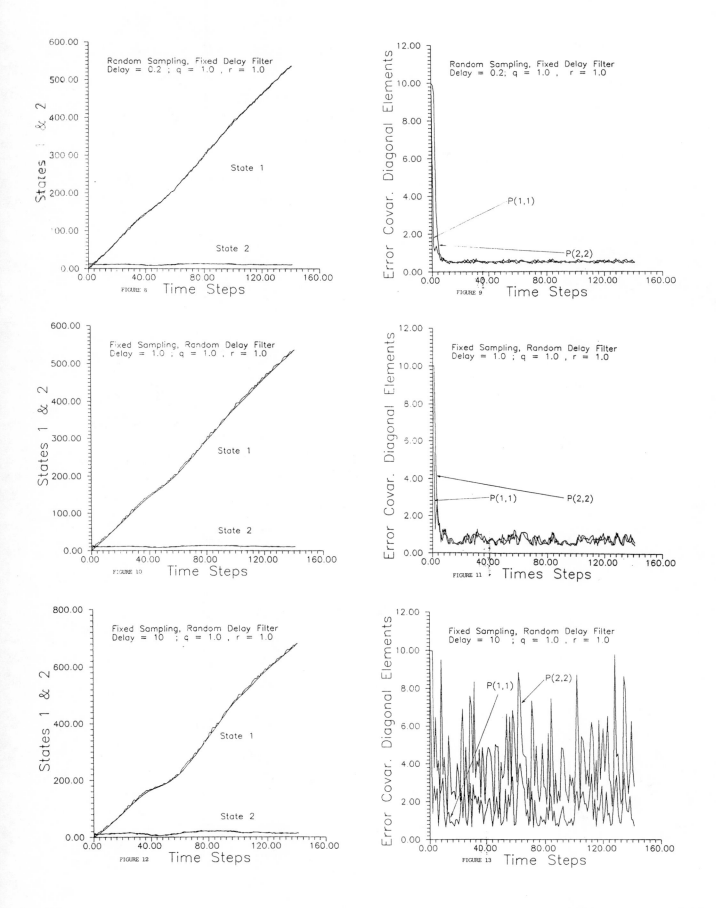

FIGURE 8

FIGURE 9

FIGURE 10

FIGURE 11

FIGURE 12

FIGURE 13

SENSOR FUSION

Volume 931

Session 3

Analysis

Chair
Charles B. Weaver
Honeywell Electro-Optics Division

Abstract only

Algorithm development for sensor fusion

Vincent C Vannicola

RADC/OCTS, Surveillance Directorate, Rome Air Development Center
Griffiss Air Force Base, NY 13441

ABSTRACT

The derivation of sensor fusion algorithms is presented with emphasis on detection and estimation of radar type targets. Theoretical expressions are developed in a form which provide the applications engineer with the fundamentals necessary for implementation of these algorithms into systems constituting distributed sensors. The expressions lend themselves to using knowledge and rule based methods so that a priori and learned information about the overall scenario can be used to reduce uncertainties and thereby efficiently direct signal energy toward optimizing system performance.

Various surveillance situations are considered and accounted for in the development of the algorithms. These include Bayesian and Neyman-Pearson detection, sequential detection, multiple target situations, estimation, colored noise such as jamming, Constant False Alarm Rate (CFAR), and multiple background estimation. Optimization of mutual information transfer through the distributed sensors is also treated. Where most investigators focus on optimizing the sensors given the fusion rule, our development explores methods for optimizing the fusion rule given the sensor criteria. Some procedures are also presented for the mutual or global optimization of both the sensors and the fusion center. The effects of band width and channel capacity constraints between sensors and fusion center are taken into account in the development. Numerical results are presented which illustrate the improvements obtained from the use of multiple sensors with various fusion rules with respect to the performance of the single sensor [1-3].

REFERENCES

I. Y. Hoballah and P. K. Varshney, <u>On the Design and Optimization of Distributed Signal Detection and Parameter Estimation Systems</u>, RADC-TR-87-130, September, 1987.

1. M. Barkat and P. Varshney, <u>On Adaptive Cell-Averaging CFAR Radar Signal Detection</u>, RADC-TR-87-160, October, 1987.

Z. Chair and P. Varshney, <u>On Hypothesis Testing in Distributed Sensor Networks</u>, RADC-TR-87-180, November, 1987.

The application of order statistic
filters in detection systems

Jafar Saniie, Kevin D. Donohue

Department of Electrical and Computer Engineering
Illinois Institute of Technology, Chicago, IL 60616

and Nihat M. Bilgutay

Department of Electrical and Computer Engineering
Drexel University, Philadelphia, PA 19104

ABSTRACT

In this paper the performance of a digital processor, referred to as the order statistic (OS) filter, is analyzed as a noncoherent processor in a detection system. The statistical description of the output of the OS filter is presented in terms of the filter parameters and the statistics of the input. The mathematical form of this description, for the case of independent and identically distributed inputs, is used to develop general input output relations of the filter. These relationships are used to indicate the critical factors that affect the performance of the OS filter, and a quantitative expression is presented to determine the rank of the OS filter necessary for optimal detection performance. It is shown that the OS filter with extreme ranks (minimum and maximum detectors) performs well in situations where a significant skewness difference exists between the different classes of input signals. As the skewness difference between the classes decreases, the performance of the OS filter with extreme ranks degrades, while the performance of the OS filter for intermediate ranks is robust over this change.

1. INTRODUCTION

The order statistic (OS) filter is a digital processor that operates on n input signals to generate output $x_{r:n}$, which is equal to the input signal value that is less than or equal to $n - r$ input values and greater than or equal to $r - 1$ input values. The OS filter can be expressed by:

$$x_{r:n} = OS_{r:n}\{x_1, x_2, x_3 \ldots x_n\} \quad \text{for } 1 \leq r \leq n \qquad (1)$$

where x_i is the unordered observed input value from the sequence of size n (window size) and r is the rank of the input from the ordered sequence that becomes the output $x_{r:n}$. This filter is the median detector when $r = (n + 1)/2$ (for n odd), the maximum detector when $r = n$ and the minimum detector when $r = 1$.

Order statistics [1] have been studied and applied in many areas of statistical theory. The application of order statistics have been discussed for image processing [2], detection of sonar signals [3], and rank-based radar detection (referred to as distribution-free detection) [4]. Order Statistics can also be applied to a radar postdetection processing technique, known as binary integration [5], which is equivalent to the OS filter. The purpose of this paper is to present the *sort function* analysis for OS filters and determine optimal rank for a general class of input signals. The sort function analysis provides insight into the properties of the

OS filter that enhance its application and optimization in the general detection problem.

In this paper the OS filter is considered as a feature extractor, and Fisher's criterion [6] for feature selection is used as an optimization criterion. The output of the OS filter associated with each rank can be considered a set of features. In the section 2 the OS filter is presented as feature extractor and in section 3 optimization via the Fisher's criterion is used to determine the optimal feature or rank. Finally, in section 4 a relationship between the skewness of the input signal distributions and the optimal rank is discussed along with robustness for certain ranks.

2. STATISTICS OF EXTRACTED FEATURES

The OS filter can be considered as an estimator of the quantile values of the input. A quantile is a value from a set of values that divide the distribution into equal probability regions. If the area under the probability density function between all pairs of consecutive points in the set are equal, these points are quantiles. In the case of 100 equal probability regions (101 points) the division points are referred to as percentiles, but in general for any number of equal probability regions the dividing points are called quantiles.

Now let $F_X(x)$ be the distribution function of the unordered independent and identically distributed input signals, and $f_X(x)$ be the corresponding density function. The expected value of the extracted feature is given by the conditional expectation:

$$E[X_{r:n}|H_i] = \mu_i \qquad (2)$$

where $X_{r:n}$ is the random variable representing the output of the OS filter and H_i is the hypothesis denoting a particular class of signals. Throughout this paper hypothesis H_0 will represent the case when no target is present, and H_1 will represent the case when a target is present.

The consistency, with which the extracted feature can be characterized by μ_i, is measured by the conditional variance:

$$E\left[(X_{r:n} - \mu_i)^2|H_i\right] = \sigma_i^2 \qquad (3)$$

Given μ_i and σ_i for both hypotheses and all possible $X_{r:n}$, the optimal feature can be determined by Fisher's criterion [6]:

$$F = \frac{(\mu_1 - \mu_0)^2}{\sigma_1^2 + \sigma_0^2} \qquad (4)$$

The maximization of F over all $X_{r:n}$ implies that the mean values of the feature for H_0 and H_1 are separated by the greatest possible distance in terms of their variances. In the following discussion the mean and variance behavior of the output of the OS filter is examined in order to obtain Equation (4) as a function of the quantile values of the input distributions.

The expression for the mean of the output of the OS filter for a specific probability distribution is given by [7]:

$$E[X_{r:n}] =$$

$$r\binom{n}{r} \int_{-\infty}^{\infty} x f_X(x) F_X^{r-1}(x) (1 - F_X(x))^{n-r} dx \qquad (5)$$

for $1 < r < n$. If $u = F_X(x)$ is substituted into Equation (5), the result is:

$$E[X_{r:n}] = \int_0^1 F_X^{-1}(u) w_{r:n}(u) du \qquad (6)$$

where $w_{r:n}(u)$ is given by:

$$w_{r:n}(u) = r\binom{n}{r} u^{r-1}(1 - u)^{n-r} \qquad \text{for } 0 \leq u \leq 1 \qquad (7)$$

and the subscript $r:n$ denotes the parameters of the associated OS filter. The factor $F_X^{-1}(u)$ in the integral of Equation (14) is the inverse distribution function. In the u domain the quantiles are easily identified as points that divide the domain into equal intervals. This follows from the fact that equal subdivisions of the u-axis correspond to the equal subdivision of the range of the distribution function.

The function in Equation (7) will be referred to as the *sort function*. The modal point of $w_{r:n}(u)$ occurs when u is equal to t given as:

$$t = \left(\frac{r-1}{n-1}\right) \qquad \text{for } 1 \leq r \leq n \qquad (8)$$

The symbol t will be used to denote the value of u where $w_{r:n}(u)$ is a maximum. An alternate notation for the sort function will use the t parameter in place of the r parameter written as $w(u)_{t:n}$. For a given n, the t values corresponding to all the possible values of r constitute a set of quantiles for the input distribution. The end points of the distribution corresponding to $u = 0$ and $u = 1$, will be referred to as the 0th quantile and the $(n-1)^{th}$ quantile, respectively.

Equation (6) reveals that the expected value of the output is the integral of the product between the sort function and the inverse distribution function. Figure 1 shows an example of the sort function with $r = 7$ and $n = 25$ superimposed on a Weibull inverse distribution function with shape parameter 1.5. Note how the sort function acts as a weighting function to emphasize a particular region of the inverse distribution function over the integration. By changing the r parameter of the sort function, the modal point can be set to emphasize different parts of the distribution.

For increasing n with t held constant the sort function is a delta sequence shifted right on the u-axis by an amount equal to t [7]. Now for a constant t as n approaches infinity Equation (6) becomes:

$$\lim_{n \to \infty} E[X_{t:n}] = \int_0^1 F_X^{-1}(u) \delta(u - t) du \qquad (9)$$

where $\delta(\cdot)$ is the delta function. At this point both n and r approach infinity, but t remains a finite ratio of n and r. Equation (9) becomes:

$$E[X_{t:\infty}] = F_X^{-1}(t) \qquad (10)$$

Since t corresponds to the endpoints of the equal subdivisions along the u-axis, the inverse distribution function of t is the $(r-1)^{th}$ quantile for a particular n satisfying Equation (8).

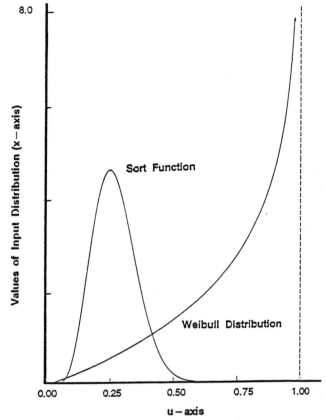

Figure 1. Sort function superimposed on inverse distribution function

It has now been shown that the OS filter is a consistent estimator of the quantile value corresponding to a constant t. For a finite n the sort function will have some dispersion about the modal point that allows the values in the neighborhood of the quantile to influence the output. This may result in a bias which must be determined to obtain the actual mean value.

Explicit results are now derived for the case when the input distribution is uniform. These results are applied later to develop a linear approximation for the general case. If the input is uniformly distributed between zero and a, the inverse distribution function becomes:

$$F_X^{-1}(u) = a u \qquad \text{for } 0 \leq u \leq 1 \qquad (11)$$

and the expected value of the output becomes:

$$E[X_{r:n}] = \frac{a r}{n + 1} \qquad \text{for } 1 \leq r \leq n \qquad (12)$$

The difference between the expected value and the modal point (same as quantile value) is given by:

$$E[X_{r:n}] - a t = \frac{a(n + 1 - 2r)}{n^2 + 1} \qquad \text{for } 1 \leq r \leq n \qquad (13)$$

Equation (13) represents the bias of the OS filter for a finite n.

Another useful quantity for determining the behavior of the output of the OS filter is the variance. For the uniform distribution between zero and a, the variance is written as:

$$M_{2,u_0} = \frac{a^2(u_0(1 - u_0)(n - 7) + 2)}{(n + 2)(n + 1)} \qquad (14)$$

where u_0 is equal to the mean value given in Equation (12).

When the input is not uniformly distributed, Equation (13) for the bias does not apply, but if the inverse distribution function in the region of emphasis of the sort function is observed to be nearly linear, then a linear approximation can be made. The value of the slope of the in the region of emphasis can be used for the value a in Equations (13) and (14). For a large enough n the slope of the inverse distribution function evaluated at the mean of the sort function, should give a good representative slope for that region.

3. FISHER'S CRITERION FOR OPTIMIZATION

The determination of the optimal quantile is equivalent to finding the optimal rank via Equation (8). In the following derivation, the optimal quantile region is determined through Fisher's criterion. For an OS filter of window size n, there exist n possible features (the n quantile values) that can be extracted. By the application of the previously developed approximations, the best quantile values can be determined from the statistical distribution of the input signals. The optimization of the filter becomes a feature selection problem.

The numerator of Fisher's criterion (without the squaring operation) in Equation (4) is expressed by Equation (6) as:

$$\mu_1 - \mu_0 = \int_0^1 \left(F_1^{-1}(u) - F_0^{-1}(u) \right) w(u)_{t:n} du \qquad (15)$$

An approximation to the integral in Equation (15) can be obtained through the use of Equations (10) and (13). From Equation (11) it is known that as n approaches infinity the distance between the mean values is given by:

$$\lim_{n \to \infty} (\mu_1 - \mu_0) = F_1^{-1}(t) - F_0^{-1}(t) \qquad (16)$$

Equation (16) is the distance between a particular quantile value for the two input signal distributions.

For a finite n, Equation (16) is a rough approximation, but the bias indicated in Equation (13) can be added to the result of Equation (16) to derive a better approximation. If the integer variable r is eliminated from the expression through a substitution with t and n, the resulting approximation for the mean distance becomes:

$$\mu_1 - \mu_0 \approx \left(F_1^{-1}(t) - F_0^{-1}(t) \right) + (a_1 - a_0) \frac{(n - 1)(1 - 2t)}{(n^2 + 1)} \qquad (17)$$

where a_1 and a_0 are representative values of the slope in the region of emphasis for $F_1^{-1}(u)$ and $F_0^{-1}(u)$ respectively. Equation (17) indicates that by increasing n the bias term

can be made arbitrarily small (for finite slope difference). Therefore, for large n the quantile distance becomes the predominate factor in mean distance of the features. The bias term becomes the predominate factor in mean distance when the quantile values are nearly equal, yet there exists a significant slope difference.

The determination of slope parameters a_i in Equation (17) is a critical factor in the quality of the approximation. Since the inverse distribution function is nondecreasing and generally well behaved (no drastic changes in the derivative over a short interval), a value of the slope along the function, centrally located in the region of emphasis of the sort function, is a reasonable choice. The slope of $F_i^{-1}(u)$ is equal to the reciprocal of the density function, evaluated at the particular x corresponding to u equal to $r/(n + 1)$ (the expected value of the sort function).

Note in Equation (17) that for increasing n, the contribution of the bias term decreases at a rate proportional to $1/n$. When either a_0 or a_1, approach infinity, then even for very large n the contribution of the bias term can be significant. This typically occurs toward the minimum or maximum values of the distribution, where the quantile distance between the two input signal distributions is usually small. Therefore, the bias term is particularly influential for the minimum or maximum detector. For large n with the OS filter operating in a range with significant quantile distance, the bias term becomes insignificant and the critical factor in these cases becomes the quantile distance.

The variance of the output feature can also be approximated in a similar manner. The sum of the variances can be obtained from Equation (14) to result in:

$$\sigma_1^2 + \sigma_0^2 \approx$$

$$\left(a_1^2 + a_0^2 \right) \frac{(1 + t(n - 1))(n - t(n - 1))(n - 7) + 2}{(n + 2)(n + 1)^3} \qquad (18)$$

Equations (17) and (18) can be applied in Equation (4) to obtain an approximation for Fisher's criterion in terms of the input distribution function, the filter window size n, and the quantile value (represented by t) at which the OS filter operates. Fisher's criterion can now be used to find the optimal r for a given set of input distributions.

4. OPTIMAL RANK AND SKEWNESS

In this section, three examples are considered to illustrate the effect of the skewness relationships between the distributions associated with H_0 and H_1 on the optimal r value. The class of signals received for H_0 will be referred to as clutter and the class of signals received for H_1 will be referred to as target. Conclusions concerning this relationship are presented in this section.

Skewness describes the deviation of the density function from symmetry about the mean value. When skewness is mentioned in quantitative terms, the measure referred to will be the moment of skewness given by:

$$s_3 = \frac{M_{3,\mu}}{\sigma^3} \qquad (19)$$

where $M_{3,\mu}$ is the third moment about the mean and σ is the standard deviation. The power ratio between the two distributions is indicated by the signal-to-clutter ratio (SCR). This value is equal to the second moment of the target over the second moment of the clutter distributions.

In the first example the clutter is more skewed than the target. This can occur with Weibull distributed clutter and chi distributed targets. The second example presents a case where the skewness of the clutter is equal to the skewness of the target. This can occur for Rayleigh distributed clutter and target. The final example presents a case where the skewness of the target is greater than that of the clutter. This can occur for a lognormal distributed target in Rayleigh distributed clutter.

4.1 Highly Skewed Clutter

When the clutter distribution is more skewed than the target distribution, the optimal r value tends toward the lower integers of the range 1 through n. This can be understood by examining the nature of the inverse distribution function for skewed distributions. Figure 2 presents the inverse distribution functions for the chi distribution with 6 degrees of freedom and the Weibull distribution with shape parameter equal to 1.5. The SCR between these two distributions is 0 dB. The chi distribution is highly symmetrical and therefore has a low skewness value. The Weibull distribution is more skewed than the chi distribution. Note how the inverse Weibull distribution functions crosses over the inverse chi distribution function for increasing u. Since it is more desirable to emphasize the regions where the smaller values of the inverse distribution of the clutter occur relative to the target, r should be chosen such that it corresponds to smaller u values.

The optimal r for a chi distributed target with moment of skewness equal to 0.31 in Weibull clutter with a moment of skewness equal to 1.07 determined by Fisher's criterion is presented in Figure 3, which is a plot of Equation 4 versus t. For the following figures, Fisher's criterion given in Equation (4) has been modified. Since it is assumed when using threshold detection that over all distribution of the target returns exhibits greater power than that of the clutter, the numerator of Equation (4) was made to reflect the sign of the difference $(\mu_1 - \mu_0)$ in the plots. Therefore, when the clutter power became greater than the target power in certain quantile regions, the criterion indicates this in a negative manner. The scale of the ordinate in Figure 3 is arbitrary since the important feature of the curve is the relative maximum point. The t values indicated by this curve are approximately 0.125 and 0.2 for SCR equal to 0 and 2 dB, respectively for $n = 25$. In this case optimal ranks obtained via Equation (8) are 4.0 and 5.8, respectively.

4.2 Target and Clutter of Same Skewness

When the target and clutter are of the same skewness value, the optimal r value tends to be in the midrange of integer values from 1 to n. Since the skewness is the same for both distributions, the major distinction between the clutter and target distributions is not any particular shape difference, but a scale difference related to the SCR. In Figure 4 two

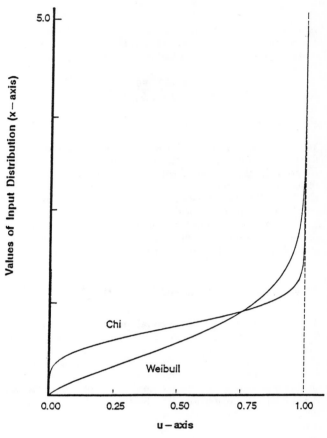

Figure 2. Target and clutter inverse distribution functions at 0 dB

Rayleigh inverse distribution functions are shown. Both distributions have a moment of skewness equal to 0.63. In the range of quantile values corresponding to the u domain between 0.2 and 0.8, two attractive features for statistical separability exist. One is a large quantile difference and the second is a small slope. The small slope implies that good variance reduction at the output of the OS filter exists.

The optimal r value for a Rayleigh distributed target in Rayleigh distributed clutter for Fisher's criterion is shown in Figure 5. Only one plot appears in this figure because at both SCR values, the curve had the same shape in both cases. This is due to the fact that a change of SCR results in a scaling of the Rayleigh distribution, but does not change the relative shape. The optimal r value is given as 18.52.

4.3 Highly Skewed Target

Finally, the case is examined where the target distribution is more skewed than the clutter. Based on the results for highly skewed clutter, it is expected that the optimal r will tend toward the higher integer values in the range 1 to n for a highly skewed target. Figure 6 indicates the relation of the inverse distribution functions for the lognormal target and Rayleigh clutter with a 0 dB SCR. The moment of skewness is equal to 2.63 for the target, and 0.63 for the clutter.

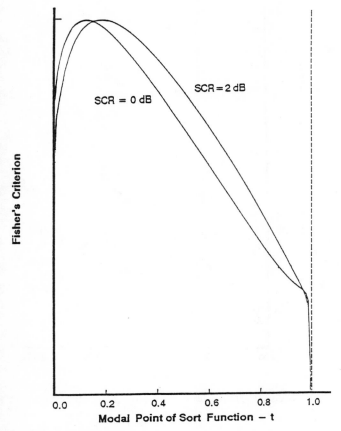

Figure 3. Fisher's criterion versus t for Chi target in Weibull clutter

The optimization curve for the Fisher's criterion is shown in Figure 7. In this case the optimal r corresponds to 25 for both SCR values. An unusual feature for the curves of Figure 7 is the presence of a local maxima in the lower quantile regions. This is due to the shape differences between the Rayleigh and lognormal distribution. Two cross over points occur between these two distributions. In Figure 6 a cross over occurs between the distribution in the regions corresponding to $u = 0.7$, and another cross over occurs, which is difficult to observe, near $u = 0.0$. This explains the local maximum near $u = 0.0$.

The relationship between the skewness of the input distributions and the optimal rank presented in this section can be useful in developing adaptive techniques for optimizing the OS filter. One major disadvantage of the optimization presented in sections 2 and 3 is that they require the entire distribution. However, if a quantitative relation can be established between the optimal (or near optimal rank) and the skewness, then the second and third central moments of distributions can be measured and the rank set accordingly.

If the skewness of the distribution is not know or is known to change over time, an alternative to an adaptive system is to determine a rank that will yield robust detection. In [4] the median detector was suggested for distribution-free detection. The idea behind this was to eliminate the problems that arise in a radar system when the assumed shape of the clutter distribution changes. If Figures 2 and 4 are compared, it is noted that, though the central quantile region is not optimal, a relatively good separation does exist

for both target-clutter combinations. If a drastic change for the input statistics occurred (as in going from the situation in Figure 2 to 4), the OS filter with rank corresponding to the intermediate quantile region would maintain a consistent performance. This is easily understood though the sort function analysis as illustrated in Figure 1, where the upper and lower quantile regions, which are generally affected by this change in the input statistics, are suppressed. Therefore, this change does not significantly degrade the performance of the decision rule.

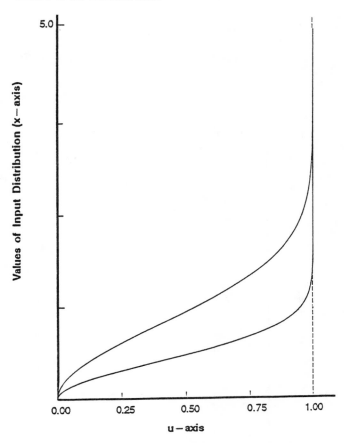

Figure 4. Rayleigh target and clutter inverse distribution functions at 6 dB

5. CONCLUSIONS

In this paper the OS filter was considered for application in a detection problem. It was shown that the parameters of the OS filter could be set to emphasize a particular quantile region where good separability between the classes of signals existed. The optimization procedure to determine the optimal rank was developed based on the Fischer's criterion. This procedure used the inverse distribution functions to find the quantile regions with the best separability. In a number of examples presented the best quantile regions could be determined by visual inspection of the inverse distribution functions.

Finally, a relationship between optimal r and the skewness of the clutter and target distributions was presented. It was shown that optimal r tends toward lower integer values when the clutter distribution was more skewed than the target distribution. When the clutter and target were of approximately the same skewness, the optimal r tended toward the midrange of integers. And when the target distribution was more skewed than the clutter distribution, the optimal r tended toward the higher integer values. These observations were made from three examples, but the results were explained in terms of the sort function analysis presented in this paper. The robust properties of the OS filters with intermediate rank values were also discussed in term of the sort function analysis.

6. ACKNOWLEDGEMENTS

The authors express their appreciation for the support of SDIO/IST funds managed by the Office of Naval Research under contract no. S400009SRB01 that made this research possible.

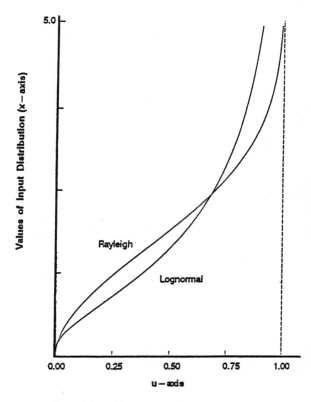

Figure 6. Target and clutter inverse distribution functions at 0 dB

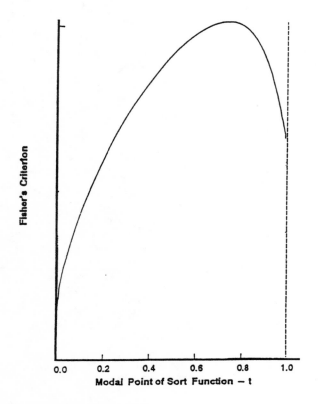

Figure 5. Fisher's criterion versus t for Rayleigh target in Rayleigh clutter

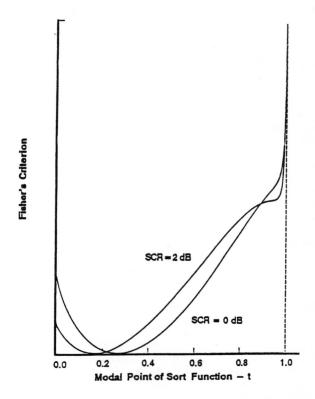

Figure 7. Fisher's criterion versus t for lognormal target in Rayleigh clutter

7. REFERENCES

[1] H.A. David, Order Statistics, John Wiley and Sons, Inc., New York (1981).

[2] Y.H. Lee and A.T. Fam, "An Edge Gradient Enhancing Adaptive Order Statistic Filter," IEEE Trans. Acoustics, Speech, and Signal Processing , vol. ASSP-35, no. 5, 680-695, May 1987.

[3] K.M Wong, and S. Chen, "Detection of Narrow-Band Sonar Signals Using Order Statistical Filters," IEEE Trans. Acoustics, Speech, and Signal Processing , vol. ASSP-35, no. 5, 597-613, May 1987.

[4] W.S. Reid, K.D. Tschetter, and R.M Johnson, "Analysis of Rank-Based Radar Detection System Operating on Real Data," The Record of IEEE International Radar Conference, Arlington, Va., 435-441, May 6-9, 1985.

[5] J.V. Harrington, "An Analysis of the Detection of Repeated Signals in Noise by Binary Integration," IRE Trans. , vol. IT-1, 1-9, March 1955.

[6] K. Fukunaga, Introduction to Statistical Pattern Recognition, Academic Press, Inc., Orlando, Fl. (1972).

[7] K.D. Donohue, Theoretical Analysis of Order Statistic Filters Applied to Detection, Doctoral Dissertation, Illinois Institute of Technology, (December 1987).

Pixel-level sensor fusion for improved object recognition

Greg Duane

Ball Aerospace Systems Division
P.O. Box 1062
Boulder, Colorado   80306

ABSTRACT

A method is proposed to exploit simultaneous, co-registered FLIR and TV images of isolated objects against relatively bland backgrounds to improve recognition of those objects. The method uses edges extracted from the TV imagery to segment objects in the FLIR imagery. A binary tree classifier is shown to perform significantly better with objects defined in this manner than with objects extracted separately from the FLIR or TV images, or with a feature-level fusion scheme which combines features of separately extracted objects. The structure of the tree indicates that the cross-segmented objects are simply ordered in feature space. An argument is presented that this sensor fusion scheme is natural in terms of the known organization of neural vision systems. Generalizations to other sensor types and fusion schemes should be considered, since it has been shown that co-registered imagery can be exploited to improve recognition at no additional computational cost.

1.   INTRODUCTION: PIXEL-LEVEL VS. SYMBOL-LEVEL SENSOR FUSION

Automated image understanding can be facilitated by considering simultaneously multiple images of the same scene produced by sensors of different types. The information provided by the various images, however, can be fused at different levels of representation. At one extreme, information can be combined at pixel-level, as the eye fuses red, green, and blue. At the other extreme, the images can be interpreted separately first, and then only the symbolic descriptions fused.

Symbol-level fusion is usually preferred when the sensors are of very different types. Within this category, choices remain as to the level of symbolic representation at which fusion will occur. Report-level fusion combines labels attached to objects in each sensory band by a separate classification scheme. Feature-level fusion[1] combines numerical or symbolic descriptions of separately segmented areas in each image before semantic labeling is attempted. For the special case of recognizing isolated monolithic objects in relatively bland backgrounds, using two dimensional signatures directly for recognition, the different modes of fusion are diagrammed in Figure 1.

Pixel-level fusion is clearly appropriate when simple quantities derived from information in different bands, such as spectral ratios, are invariant to scene conditions, such as illumination. When the images are of very different types, it is usually not possible to construct analogous invariant quantities. Another argument against the pixel-level fusion approach is that human beings do not readily establish or utilize a point-to-point correspondence between images from different sensor types. However, it is the point of this paper that images from different sensor types can still be efficaciously fused at pixel-level. The fusion is based on the observation that the presence of certain substructures, such as edges, in any one image, can almost always be taken to indicate their physical presence in the scene.

In section II, a simple scheme is presented to combine segmentations of objects appearing simultaneously in co-registered TV and forward-looking infrared (FLIR) images. In section III, an experiment is described which shows that a crude object classifier provides better classifications using the multi-sensor segmentations than using either the single band segmentations or the combined sets of features extracted from the single band segmentations separately. In section IV, an argument is presented that the sensor fusion scheme described is natural in terms of the organization of current neural vision models. Section V summarizes conclusions and suggests possible generalizations.

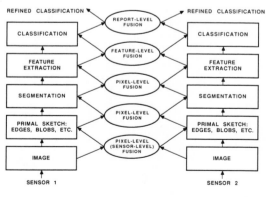

Figure 1. Sensor fusion at various levels of representation.

2. MULTI-SENSOR SEGMENTATION OF OBJECTS IN FLIR AND TV IMAGERY

The term pixel-level fusion is taken to mean fusion at any level of abstraction for which image co-registration, accurate to within a pixel, is required to obtain accurate results through the fusion algorithms. At the lowest level of representation, consisting of raw pixel data, it appears unlikely that FLIR and TV images can be efficaciously fused. Indeed, pixel-level TV image data is essentially useless for recognition because of the large variance in TV grey levels (reflected intensities) with illumination and other scene conditions. FLIR grey levels are somewhat more informative quantities, since certain objects are typically hot, etc., but cannot be meaningfully combined with TV data.

At the next level of representation, though, fusion appears somewhat more promising. This level, corresponding roughly to Marr's notion of a primal sketch[2] in the more comprehensive vision blackboard scheme, consists of entities in the image plane such as edges and blobs, as yet meaningless, but to which physical meaning may later be attached. Primal structures, in one image, if sufficiently distinct, suggest the presence of corresponding structures in the other image. (At a lower level of resolution, this observation is the basis for fusing sets of detections, prior to segmentation or recognition.) In the special case of isolated objects against a bland background, primal structures derived from one image can be simply used to help distinguish object pixels from background pixels in the other image, and hence to improve the final segmentation. While feature-level fusion ignores the relative physical positions of corresponding objects as depicted in the two images, pixel-level fusion makes essential use of this information (Figure 2).

A crude approach is to take the set of pixels which constitute an object to be the union of the set derived from primal structures in one image with the set derived from primal structures in the other image. This will be especially effective if the performance of the subsequent classification algorithm suffers less from missegmenting background as object than missegmenting object as background (e.g., if classification depends on grey-level moments of bright objects against dark backgrounds). Under other circumstances, it might be preferable to require that a pixel be designated as object in both images before it is assigned to the object in the final segmentation. A possible generalization would be to consider "fuzzy segmentations" in which each pixel is assigned a probability of belonging to the object in each image and then to combine probabilities accordingly. Still a fourth approach would be to construct a knowledge-based system which would consider the particular description of each primal structure before deciding

how to incorporate it into the final segmentation in either image. Heuristics would be based on tentative classifications of structures or on domain-independent knowledge such as that used by Nazif and Levine[3] for single-band imagery.

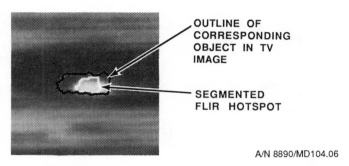

OUTLINE OF CORRESPONDING OBJECT IN TV IMAGE

SEGMENTED FLIR HOTSPOT

A/N 8890/MD104.06

FLIR IMAGE

Figure 2. Pixel-level fusion makes use of relative positions of objects extracted separately from FLIR and TV; feature-level fusion does not.

The multi-sensor segmentation approach adopted in this work is somewhat simpler. It is noted that the temperature edges which define object boundaries in FLIR imagery, are characteristically less distinct than are reflected intensity edges which define object boundaries in TV imagery. Conversely, while objects in FLIR imagery typically display temperature patterns which are invariant to scene conditions, objects in TV imagery display intensity patterns which vary widely. In the present experimental context, TV images, in fact, contribute only shape information to the recognition process. Therefore, as a first step towards the construction of a more comprehensive multi-sensor segmentation algorithm, we take segmentation information from the TV imagery alone. That is, an edge-based segmentation algorithm, applied to TV images, is used to delineate objects which are filled in with photometric data from the corresponding FLIR images for subsequent feature extraction and classification.

3. AN EXPERIMENT IN MULTI-SENSOR OBJECT RECOGNITION: EDGES FROM TV; REGIONS FROM FLIR

The proposed multi-sensor segmentation scheme was applied to a data base of simultaneous FLIR and TV images of military vehicles against natural backgrounds, acquired aerially from distances of 2.5-5 km. The images were co-registered manually, by fitting the parameters of an affine transformation to match from three to six pairs of corresponding points, selected by the operator, in each image pair.

The edge-based segmenter employed to delineate objects in both the FLIR and TV images was the well known algorithm due to Perkins.[4] The algorithm iteratively extends edges, obtained by thinning and thresholding edges in a gradient image, by successively adding increasing numbers of pixels to the endpoints of the edges. (The algorithm employed in this work differs from Perkins' original algorithm in that edges are only extended from their endpoints.) After each iteration, only those extension pixels are retained which bound new regions. After a small number of iterations, extraneous edges are discarded, leaving only closed edges (e.g., Figure 3). After defining very large regions in the edge-linked output as background, clusters of adjacent non-background regions were merged, to yield segmentations of candidate objects in the scene.

Thresholds were set conservatively so that only true edges in both images contribute to the segmentations. Furthermore, since the purpose of this experiment was to compare different modes of segmentation, only those objects were retained which were segmented separately in both the FLIR and corresponding TV images. (In a real application, segmentation would be preceded by a prescreening or detection stage, to focus attention on small areas of interest, but we omit this stage.)

Four modes of object recognition were compared. First, a tree classifier was constructed to distinguish among the three classes: tracked vehicles, trucks, and clutter objects, using feature data extracted from object regions segmented separately (Figure 4) in the FLIR image. Then such a classifier was constructed for objects segmented separately (Figure 5) in the TV image. Thirdly, a classifier was constructed to use a feature set which was the union of feature sets extracted separately from separately segmented objects in the TV and FLIR images. This is the feature-level fusion scheme. Finally, a classifier was constructed to use features extracted from the FLIR images in object regions delineated by the boundaries of corresponding objects in the TV images (Figure 6).

The construction of the classifier followed the procedure described in reference 5 for growing and pruning binary classification trees. A uniformity test is applied to a training data set to select a particular feature and a threshold value to split the data set into two subsets which are maximally "pure." The optimal choices are defined to be those which minimize the function:

$$\sum_{i \in \text{CLASSES}} (-P_{i\,|left} \log P_{i\,|left}) + \sum_{i \in \text{CLASSES}} (-P_{i\,|right} \log P_{i\,|right})$$

(1)

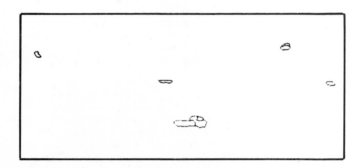

Figure 3. Output of Perkins edge-linking algorithm.

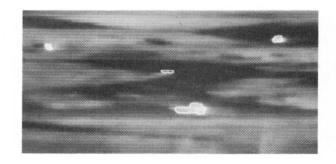

Figure 4. Object/background segmentation of FLIR image.

Figure 5. Object/background segmentation of TV image.

Figure 6. TV-based segmentation of FLIR imagery.

over all possible splits into "left" and "right" subsets. The procedure is repeated for each of these two subsets. A stopping criterion is applied to determine when a data subset is sufficiently "pure" as to require no further splitting. In this experiment, the stopping criterion was that all samples in a data subset are of the same class. Finally, the tree is pruned, using a separate data set, by merging branches which do not

sufficiently differentiate object classes in this new data set. The pruning algorithm acts to minimize a cost function which is a linear combination of tree size and misclassification rates weighted by relative costs. In this experiment, all types of misclassification were weighted equally.

A binary tree classifier was chosen for this experiment in part because, unlike conventional Bayesian or KNN classifiers, the tree classifier makes no assumptions about the structure of the data set in feature space. Such assumptions would normally be incorporated in the coefficients used to compare different features so as to define a metric in feature space in the conventional schemes. Because structure is not assumed, it is in fact discovered by the tree growing process. Thus, tree structure, as well as classification performance statistics, will be used to evaluate the results of this experiment.

Confusion matrices for the various recognition schemes are shown in Figure 7, together with the structures of the associated classifier trees. It is apparent that pixel-level fusion for multi-sensor segmentation gives performance superior to that obtained with the other three schemes (Figure 8), reducing the error rte by a factor of 2 or 3. The effectiveness of this classification

(ROWS ARE TRUE CLASSES; COLUMNS ARE COMPUTED CLASSES)

14 OFF-DIAGONAL OBJECTS

	TRACKED	TRUCK	CLUTTER
TRACKED	13	1	4
TRUCK	7	2	0
CLUTTER	2	0	5

FLIR ALONE

13 OFF-DIAGONAL OBJECTS

	TRACKED	TRUCK	CLUTTER
TRACKED	18	0	0
TRUCK	9	0	0
CLUTTER	4	0	3

TV ALONE

5 OFF-DIAGONAL OBJECTS

	TRACKED	TRUCK	CLUTTER
TRACKED	16	1	1
TRUCK	1	8	0
CLUTTER	2	0	5

PIXEL-LEVEL FUSION

14 OFF-DIAGONAL OBJECTS

	TRACKED	TRUCK	CLUTTER
TRACKED	13	1	4
TRUCK	7	2	0
CLUTTER	2	0	5

FEATURE-LEVEL FUSION

A/N 8890/MD104.13

Figure 7. Confusion matrices and associated classifier tree structures for the various recognition schemes.

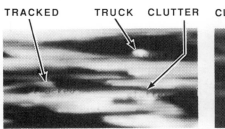

FLIR IMAGE: TRUE CLASSES

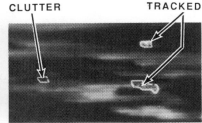

FLIR CLASSIFICATIONS

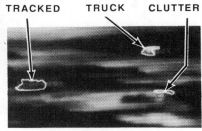

PIXEL-LEVEL FUSION

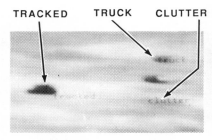

TV IMAGE: TRUE CLASSES

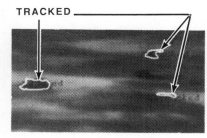

TV CLASSIFICATIONS

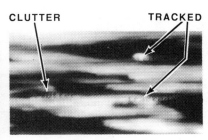

FEATURE-LEVEL FUSION

A/N 8890/MD104.12

Figure 8. Classification results for typical imagery.

scheme in distinguishing among vehicle types is particularly striking. Equally important is the relative simplicity of the tree structure in this case. (For the TV case, although the tree is simple, classification performance is abyssmal. This is interpreted to mean that the tree-growing algorithm is incapable of finding a classifier tree to distinguish among vehicle types from their silhouettes alone, probably since in our data vehicles in the TV imagery are uniformly dark.) These results suggest that objects formed by filling in TV outlines with photometric data from FLIR are simply ordered in feature space.

4. A BIOLOGICAL ANALOGUE: BOUNDARY CONTOUR SYSTEM AND FEATURE CONTOUR SYSTEM

In answer to the argument that the pixel-level fusion scheme proposed in this work is unnatural, it is suggested that the proposed definition of roles for TV and FLIR imagery is strongly reminiscent of a particular division of labor within biological vision systems. Reference is made to S. Grossberg's analysis[6] of numerous visual phenomena to conclude that two very distinct neural subsystems contribute to segmentation processes (Figure 9). The first of these subsystems, the boundary contour system, detects and links edges which outline regions that are filled in by color and brightness signals from the second subsystem, the feature contour system. If the boundary contour system is deactivated through image stabilization, filling-in proceeds unchecked and adjacent regions, of different color and brightness, blend completely. Boundary contour signals are completely invisible. The boundary contour system is insensitive to direction of contrast and relatively insensitive to color,[7] while the feature contour system is sensitive to direction of contrast and generates perceived colors. Both systems are separately linked to a higher level object recognition system.

It is suggested that the structure of lower level neural vision systems be consid-

ered, because we seek to develop not artificial intelligence but artificial perception. While human beings do not see infrared, but use higher level faculties to combine visible imagery with infrared imagery rendered as visible imagery, the desired FLIR/TV-based object recognition system should function less awkwardly. If the edge-linking algorithm used in this work is analogized to the boundary contour system, applying this "boundary contour system" to TV imagery while applying a "feature contour system" to the FLIR imagery, in view of the distinctness of the analogue biological systems, would appear to be a particularly natural scheme for low-level sensor fusion (Figure 10).

5. CONCLUSIONS AND POSSIBLE GENERALIZATIONS

It is hoped that the conclusions to be drawn from this work as to the utility of pixel-level fusion of imagery acquired from TV and FLIR sensors can be generalized to other combinations of sensor types. In particular, it is thought that the fusion scheme proposed might be advantageously applied to LADAR/FLIR image pairs. Because range edges are commonly very sharp, LADAR signals might be used to delineate regions that would be assigned temperature patterns derived from FLIR imagery. Region-based segmentation of the LADAR imagery could also be used for this purpose. A surface model augmented with temperature information in this manner might provide valuable recognition clues.

A second generalization is to use the multi-sensor segmentation approach within a structural classification scheme. With better sensors, or at shorter ranges, objects can be recognized through relationships among their parts, rather than considering them as monolithic wholes. Multi-sensor segmentation might be used in this context to delineate object parts.

The fusion approach described thus far combines data at the highest level of representation which still requires pixel-to-pixel correspondence (Figure 1). At the level just below this one, numerous possibilities exist

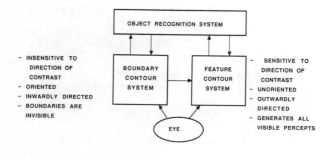

Figure 9. Two completely distinct neural subsystems contribute to the segmentation of visible objects.

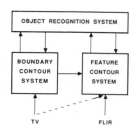

Figure 10. Conjectured neural architecture for diurnal beings equipped with FLIR and TV sensors.

for combining primal structures in both images to obtain a novel segmentation rather than simply transposing a segmentation from one image to the other. It is not known whether more complex strategies of this sort will be very effective.

This work has determined that pixel-level fusion of different types of images can be beneficial. While a quantitative assessment would be inappropriate, significant reductions in error rate were shown to occur. It is concluded that, if coregistered imagery is available, it can be algorithmically exploited in typical scenarios, to significantly improve object recognition rates at no additional computational expense.

6. ACKNOWLEDGEMENT

The author wishes to thank James Oligschlaeger for suggesting the use of a binary tree classifier in the experiment described in this paper, and for implementing the classifier.

7. REFERENCES

1. W.G. Pemberton, et. al. "An Overview of ATR Fusion Techniques" presented at Tri-Service Data Fusion Symposium (1987).

2. D. Marr, Vision, Freeman, San Francisco (1982).

3. A. Nazif and M. Levine, "Low Level Image Segmentation: An Expert System," IEEE Trans. on Pattern Analysis and Machine Intelligence, Vol. PAMI-6, No. 5, 555-577.

4. W.A. Perkins, "Area Segmentation of Images Using Edge Points," IEEE Trans. on Pattern Analysis and Machine Intelligence," Vol. 2, No. 1 (1980).

5. L. Breiman et. al. Classification and Regression Trees, Wadsworth, Belmont, Ca. (1984).

6. S. Grossberg, "Cortical Dynamics of Three-Dimensional Form, Color, and Brightness Perception: I. Monocular Theory," Perception and Psychophysics, 41(2), 87-116 (1987).

7. M.S. Livingstone, letter to Scientific American, 258(3), pg. 6, (1988).

SENSOR AND INFORMATION FUSION
FROM KNOWLEDGE-BASED CONSTRAINTS

Allen R. Hanson, Edward M. Riseman
Computer & Information Science Department
University of Massachusetts[1]
Amherst, MA 01003

Thomas D. Williams
Amerinex Artificial Intelligence, Inc.
274 N. Pleasant Street
Amherst, MA 01002

ABSTRACT

A constraint-based approach to uniformly combining information from multiple representations and sources of sensory data is described. The approach is important to research in intermediate grouping, knowledge-based model matching, and information fusion. The techniques presented extend the capabilities of an earlier system that applied constraints to attributes of single types of extracted image events called tokens. Relational measures are defined between symbolic tokens so that sets of tokens across representations can be selected and grouped on the basis of constraint functions applied to these relational measures.

Since typical low-level representations involve hundreds or thousands of tokens in each representation, even binary relational measures can involve very large numbers of token pairs. Control strategies for ordering and filtering tokens, based upon constraints on token attributes and token relationships, can be formed to reduce the computation involved in producing token aggregations. The system is demonstrated using region and line data and an associated set of relational measures. The approach can be naturally extended to include tokens extracted from motion, stereo, and range data.

1 Introduction

A major problem confronting vision systems which use multiple sensors, or which generate multiple low-level descriptions from image data, is the coherent and consistent integration of information contained in the multiple representations. Most vision systems utilize only one type of sensory data (e.g., visible light, SAR, IR, range) and only one type of low-level process producing a single type of extracted image event, e.g., regions of a region segmentation. However, after many years of computer vision research [HAN78a,b,HAN87,RIS87] it is clear that such systems are fundamentally limited by their restricted and unreliable view of the image data, and consequently their performance must suffer by the degree to which the image descriptions fail to support the system's goals.

More recently, multiple sensors have become more readily available and many algorithms have been developed for processing each type of sensor data. For example, depth maps can obtained directly from laser range data, and indirectly from motion and stereo algorithms that are applied to pairs and sequences of images, respectively. It has also become evident that each low-level process extracts only partial descriptions of the underlying image structure, and that there is a great deal of redundancy

which can be profitably exploited across these descriptions. Consequently, the need is becoming more acute for computer vision systems to fuse the information extracted by different types of low-level vision algorithms into more coherent descriptions. Maximum reliability can only be achieved through processes that can integrate information represented in widely varying forms.

To be somewhat more specific, consider the interpretation of a road scene. The formation of a 'road' hypothesis should not be based on any single type of extracted image event (e.g. regions), but rather on an aggregation of multiple types of events (e.g. lines, regions, and surfaces) that have specific relationships to each other and which contribute to the support of the road structure. For example, one might like to find a homogeneous region of an expected intensity and color, bounded by two converging straight lines, and approximately covered by a horizontal planar surface.

Of course the reader should not be misled into an oversimplified view of the problem; there are extremely difficult low-level issues to be dealt with, such as the instability of segmentation algorithms that leads to unpredictable fragmentation of lines, regions, and surfaces [BEV87,KOH83,NAG82], and inconsistencies between the elements extracted in these representations. These are problems that are implicit in the nature of the problem of integrating unreliable information and will be true of all approaches, not just the one presented here. Our view is that to fully integrate multiple representations there will need to be complex grouping strategies that utilize the techniques presented here as part of a knowledge-directed interpretation process [BOL87,DRA87a,b,HAN87,REY87b].

In this paper, we take the view that information fusion can be accomplished during later stages of the interpretation process, rather than when the image events are first extracted (e.g. by attempting to directly integrate region and line algorithms). We also believe this will avoid some of the severe ambiguity problems encountered when performing interpretation on the independent representations prior to information fusion. Our approach to fusion will be illustrated here by extending a constraint-based object hypothesis system [RIS87] to operate over multiple token types, in this case regions and lines extracted from the image data. The construction of the region and edge representation makes use of two low-level algorithms: a local histogram-based region segmentation algorithm [BEV87,KOH83,NAG82] and a straight line extraction algorithm [BUR86]. Examples of these two processes are shown in Figure 5b,c. When surface elements extracted from depth maps are available, they can be aggregated with regions and lines. The techniques could also be easily extended to include fusion of information from textured areas, corners, volumes, and generally any other token abstracted from the same or other sensory sources.

There have been a few attempts to integrate results from multiple low-level processes operating on one or more sensory sources

---

[1]This work has been supported by the Defense Advanced Research Projects Agency (DOD), ARPA Order No. N00014-82-K-0464, Air Force Office of Scientific Research under contract F49620-83-C-0099, and the National Science Foundation under contract DCR-8500332.

[HAN78b,KOH83,NAS83]. In the past, efforts to combine multiple processes operating on visual data have typically involved the integration of line and region data, which are the two most common types of low-level algorithms employed. More recently, there has been an increasing number of efforts to combine range and visual data [ARK87,BES85,SHA86]. Shafer and Thorpe [SHA86] developed a blackboard system for the CMU NAVLAB mobile vehicle; in this system, range data and visual data are independently processed and combined during the interpretation process. On the other hand, Nandhakumar and Agarwal [NAN87] combined the processing of infrared and visible light images with computational models of the image generation process to improve the results beyond that achievable by either process alone.

2 Background

2.1 The Intermediate Symbolic Representation (ISR)

The most general model of image interpretation involves the construction of a symbolic three-dimensional description of a scene from an image or set of images; a related goal is the identification of a specific 'target' object from background clutter. It is generally accepted that a computer vision system must perform a variety of transformations of the data during this interpretation process. Consequently, at a coarse conceptual level, the VISIONS image understanding system is organized into three levels of processing: low, intermediate, and high as shown in Figure 1. Currently, the low-level, or segmentation, processes output a symbolic representation of the data in the form of regions and lines. Attributes, such as color, texture, location, size, shape, and orientation, are then calculated for each region or line. Interpretation processes use knowledge of the objects in the domain to control a set of intermediate-level processes for generating initial object hypotheses and reorganizing the low-level data. World (domain) knowledge is then responsible for resolving these hypotheses data into a consistent model of the scene.

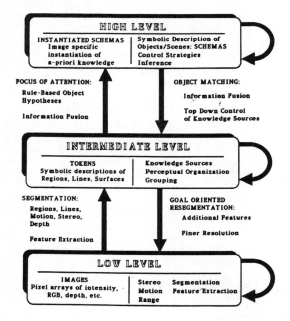

Figure 1. The VISIONS System: Processing and control across multiple levels of representation are depicted in this system overview.

One of the key abstractions is the transformation of pixels, or more generally arrays of sensory data, into image events which can be named and accessed by their properties. We refer to the symbolic representation of an extracted image event, such as a particular region, line, curve, rectangle, or surface, etc., as a *token*; attributes are associated with tokens, and tokens participate in relationships with other tokens. Note that tokens may be defined in terms of other, more primitive tokens, as in the case of a rectangle. A *tokenset* is a collection (set) of tokens of the same type (e.g. region tokens). Tokensets, tokens, token attributes, and relations between tokens are all organized into a type of relational database called the "intermediate symbolic representation" (ISR). The ISR allows flexible associative access to tokens and serves as the communication interface between the low-level descriptive processes and the high-level interpretive processes in the VISIONS system. In general, the only requirement for placing a new type of low-level token into the ISR is that each primitive element of that data type must have a symbolic name (e.g., region–240, surface–38, corner–46) and a non-empty set of attribute-value pairs. It is the values of the token attributes, and, as we shall present here, the relational information between tokens, that provide the basis for initial interpretation processes.

3 Constraints on Tokens of a Single Type

3.1 Constraint Functions on Token Attributes

A simple type of knowledge source for generating hypotheses of object class labels for particular regions has been under development in the VISIONS environment for some time [BEL86,HAN87a] [REY87a,RIS84,WEY83,WIL81]. The general idea is to develop a mapping from a region token and its attributes onto an object label hypothesis for the region, e.g. 'grass'. We note that the VISIONS system operates primarily in the outdoor scene domain, but the general techniques developed below are applicable to most scene domains and sensor modalities. The mapping was accomplished by defining constraints on the range of an attribute from a sample set of the object and a constraint function which mapped from the region attribute into a weighted 'vote' or 'score' for the object label (see Figure 2). Compound constraints were defined as (possibly recursive) combinations of the output of a set of these 'simple' constraint functions. Region features such as region color, texture, shape, size, image location, and relative location to other objects were used. More recently, the approach has been extended to lines, using attributes such as length, orientation, contrast, width, etc. While no single constraint on the features of a region or line can ever be totally reliable, the combined evidence from many such constraints often imply the correct interpretation of a token; for example, in a rank ordering of a set of regions on the basis of the final 'score', the region-object label association for those regions near the top of the list is often correct. In many cases, it is possible to define constraints which provide evidence, in the Dempster-Shafer sense, for and against the semantically relevant concepts representing the domain knowledge [REY85,REY87a].

Rather than viewing the application of the constraint set through the constraint function as a classification process in the pattern recognition sense, the rank- ordered output can be used as an unreliable set of hypotheses and used to trigger focus of attention mechanisms in an artifical intelligence sense [HAN78b]. They are used by other more complex knowledge-based processes in a hypothesize and verify control structure [DRA87a,b,HAN87a] [WEY86].

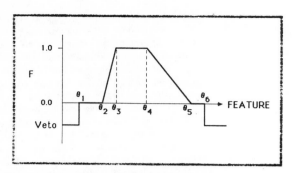

Figure 2. Structure of a simple constraint function as a piecewise linear function F mapping an image feature measurement into support for an object label hypothesis. It is specified via 6 points $\{\theta_1, i = 1, \ldots 6\}$.

A simple constraint function (which hereafter will often be referred to simply as a constraint) is a function F applied to the kth attribute (or feature) of the jth token of type T. Thus, if the kth attribute of line tokens (type L) is length and L_{jk} is the length of the jth line token, then $F(L_{jk})$ would be the response of the constraint function when applied to the line length of token j. A variety of forms for the function F have been employed, with no appreciable difference in the results. The first was an extended real-valued piecewise-linear function F

$$F(T_{jk}) \in \{[0,1] \cup VETO\} \qquad (1)$$

specified by six points in the feature range $\{\theta_i, i = 1, \ldots 6\}$ as shown in Figure 2 [RIS87,WEY83,WEY86]. The simplicity of this approach is that F in this form could be compactly stored as a 6-tuple, or sets of 6-tuples.

Compound constraints are a hierarchical collection of simple and compound constraints with an arithmetic or logical combination function for collecting the individual responses into a single response. For discussions of variations on compound constraints see [HAN87a,KIT86,RIS84,RIS87,WEY86]. In some experiments, the top-level compound constraint for an object was

structured as a combination of five other compound constraints (as shown in Figure 3) to represent color, texture, size, shape, and location constraints, each of which was composed of a set of simple constraints [RIS87].

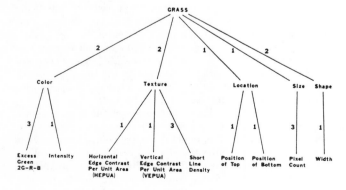

Figure 3: The structure of a compound constraint for grass showing the five component constraints defined on region attributes.

3.2 Relational Similarity Constraints on Tokens of the Same Type

When dealing with unreliable segmentation processes, forming aggregations of tokens is usually necessary since the set of tokens need to be grouped and reorganized in order to match an object model [DRA87b,HAN87a,HAN87b,REY87b]. The basis of this grouping usually involves not only the attributes of tokens, but also the relations between tokens. The constraints described in the previous section are unary, since they accept a single token attribute as input and return a value that can be viewed as a confidence or rating for the hypothesized object. The highest ranked of these hypotheses can serve as a partial (and probably errorful) interpretation of the original image. However, it is clear that constraints on relationships between tokens are also fundamental to object recognition. They can be handled in much the same way by defining constraints on relational measures (a function which quantifies a particular relation) between pairs of tokens. In this case, the response of a constraint function specifies the degree to which the constraint on a relational measure is satisfied.

Let us consider as a simple example a line token set and a relational measure defined by the simple absolute difference of their orientation attribute. Given a specific line token (e.g. L-435), all other tokens can be rank-ordered relative to L-435 by a constraint function applied to the orientation difference. Figure 4 shows two different constraints F_1 and F_2 for processing line tokens relative to a given line token. F_1 gives a maximum response of 1 for all tokens that are within 10° of L-435, a linear decreasing response from 10° to 30° and a veto response beyond 45°. The effect of applying F_1 is to rank equally all lines that are very similar in orientation to that of L-435; beyond 10° they are ordered based on their relative orientation. Lines whose relative orientation is greater than 45° away from that of L-435 are excluded. The constraint embodied in F_2 results in the selection of all line tokens that are within 5° of being orthogonal to L-435. In effect constraint F_2 defines a relation on approximately orthogonal pairs of lines that have L-435 as a member; the relation is defined to have a value of True for all pairs where $F_2(L_{435,k}; L_{j,k}) = 1$. In the case of constraint F_1, the relational measure for those lines not vetoed is mapped into a response which can be used to coarsely rank order the line pairs in terms of how 'parallel' they are. Note that applying a threshold to the respnses produces a true relation. Either of the subsets resulting from application of these constraints could be followed by a token attribute constraint for ordering or filtering the remaining lines on other attributes such as location, contrast, length, etc. in absolute terms or relative to L-435. In the next section we show how this general idea can be extended to tokens of different types, resulting in a fusion of the information from the two different sources.

4 Integrating Representations via Construction of Token Aggregates

The fundamental problem that is being addressed in this paper is the integration of multiple low-level representations into the interpretation process. While the approach presented here offers only one type of information fusion mechanism and deals with only some of the most general levels of the information fusion problem, there are several important advantages. First, it offers an entirely modular and natural method for incorporating additional processes and representations as a vision system undergoes incremental development; in particular, existing low-level representations do not have to be modified in any way. Secondly, the integration is accomplished at the intermediate group-

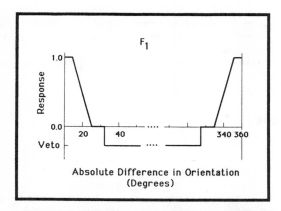

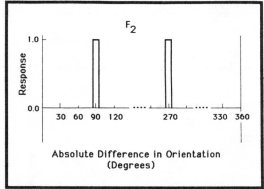

Figure 4. Example Constraint Functions on Relational Measures. The structure of two constraint functions for relating one line token to another is shown. The constraint is applied to the difference in orientation of two line tokens. (a) F_1 equally ranks all lines whose orientation is within $\pm 10°$ of the given line; the response falls off as a function of the difference in orientation. (b) F_2 equally ranks all line tokens that are within $\pm 5°$ of being orthogonal to the given token.

ing and/or interpretation levels through constraints which relate entities in the independent representations. Since there is no direct interaction of the processes which initially create the tokens, the mistakes of one low-level process will not affect the output of the other low-level processes. If a sufficient body of consistent information exists in several representations, then low-level mistakes in a given representation may be detected and either ignored or corrected, as opposed to integrating partially erroneous data in some form, such as a least-squares optimization process. Third, this approach is an extension of an approach applied to token attributes that has already proven to be reasonably effective on very complex natural scenes [HAN86,HAN87a,RIS84]. Finally, the techniques can be used as part of general intermediate grouping processes. The grouping can be viewed as knowledge-directed (e.g. via a model of an object), or it could be viewed as a data-directed token aggregation process whose goal is to extract interesting structures of a priori importance [REY87b].

4.1 The Formation of Aggregations of Tokens

Relational constraints are defined as a real-valued function on a relational measure defined over a token set. A *relational measure* $M(T_{j_1 k_1} \dots T_{j_m k_m})$ is a function of the attributes of multiple

tokens, possibly of different types. As we have already discussed, for tokens of the same type there is an implicit binary relational measure in that the same scalar-valued attribute of any two tokens can be compared by their similarity or difference. For example, given a specific region, similarity relational measures be used to compute the distance between feature centroids as well as the difference in mean intensities between the given region and all other regions. Once such a scalar relational measure exists, a constraint function can be applied to the relational measure to produce a response that represents the degree to which the relational constraint is satisfied.

In order to compare tokens of different types a relational measure must be defined between each pair of token types. Therefore, if new sensors are added, and new sensory events are extracted, relational measures must be defined between the existing token types and the new token types. This will allow information from new representations to be integrated. (In this paper, only binary measures between region and line tokens are developed.) The constraint functions on relational measures can then be applied to sets of tokens across the multiple representations in order to group tokens into aggregations. For example, relational measures between line tokens and region tokens can be defined, such as the degree of intersection between tokens; a constraint on this relational measure could then select, for each region, all lines that are sufficiently interior to the region. The reader should note that when a relational constraint is used to filter the token tuples, into disjoint sets, the result is a relation.

4.2 Region-Line Relational Measures

We now present a specific set of relational measures to provide a computational method of relating regions and lines. Relational constraints on these relational measures will then be used to implement the following relations between regions and lines:

- BOUNDING lines – those lines associated with a region boundary;

- INTERIOR lines – those lines interior to a region; and

- OTHER lines – those lines which intersect a region, but are neither bounding nor interior to the region.

The relational measures chosen are based on the intersection of sets of pixels. Only lines which intersect a given region are of interest. We will represent lines by the subset of pixels (called the line-support-set,) comprising that portion of the intensity surface that led to the extraction of the line [BUR86]; thus, the line-support-sets of pixels can be expected to overlap the regions that they bound or are interior to. Consequently, INTERSECTION (defined in the usual way on pixel subsets) becomes a natural relation which can be used as a filtering constraint to select a subset of tokens. In the following discussion, however, three other relational measures will be defined and used to implement the three relations mentioned above: "interior-line-percentage", "region-perimeter-percentage", and "line-boundary-percentage".

The first relational measure, "interior-line-percentage", is the ratio of line area interior to the region to total line area. The interior-line-percentage measure discriminates lines that are entirely INTERIOR from BOUNDING lines, whose line-support set will lie partially outside the region. An INTERIOR line will have a value of 100% for this relational measure, indicating that the line-support-set is completely contained by the region. An ideal BOUNDING line with a symmetric line-support-set of pixels lying exactly on the region boundary would have half its pixels in the region and a value of 50% for its interior-line-percentage.

The other two relational measures can be used to discriminate BOUNDING lines from INTERIOR lines. The natural duality between regions and their boundary lines can be exploited in a straightforward manner to indicate how much of a region boundary or a line is covered by the other. "Region-perimeter-percentage" measures the fraction of a region boundary made up of one line and is defined to be the ratio of the intersection of the region perimeter pixels and the line-support-set to the length of the region perimeter. "Line-boundary-percentage" measures the fraction of a line contributing to the region boundary and is defined to be the ratio of the intersection of the line-support-set and the region perimeter to the total line length in pixels. Ideally, a line which lies approximately on a region boundary will have a high value of line-length-percentage since the region boundary will cover much of the line. The same is true of region-perimeter-percentage although a single line will be expected to cover a smaller portion of the entire region boundary.

4.3 Relational Constraints

Relational constraints are used as the final step in the formation of aggregations from multiple representations. The relational measures presented in the preceding sections provide the basis for defining these relational constraints.

A relational constraint function for lines and regions can be specified for each relational scalar measure that has been defined (in the same manner that token attribute constraints are defined). Thus, a simple constraint can be specified for each of interior-line-percentage, line-length-percentage, and region-perimeter-percentage measures; note that any of these simple constraints may be omitted. A combination function can then be defined for combining the output of the set of simple constraints into a compound relational constraint.

The form of the function that combines the simple constraints is not critical, and in this paper we will use the same simple piecewise-linear function described earlier with a range of $\{[0, 1] \cup VETO\}$. The VETO range(s) serves only as a first filter for selecting or removing candidates for processing, in the sense that the vetoed tokens do not satisfy the constraint. This does in fact, define a relation over the token sets, but the remaining non-vetoed tokens in the relation still have the graded response from the constraint function, which can be used for ranking or further filtering of token pairs to produce a more restricted subset of tokens.

4.4 Controlling the Formation of Token Aggregations

The aggregation of tokens via relational constraints must, of course, contend with the combinatorics of the large number of image tokens whose relationships must be examined. The concept of *focus-of-attention* becomes important when one considers that the representations being used typically involve 2,000 to 10,000 lines and 200 to 1,000 regions. Thus, there are potentially 400,000 to 10 million line-region pairs which could be related, and these numbers become much larger as additional representations, multiple sensory sources, finer image resolutions, or larger images are considered.

The order and manner in which attribute constraints and relational constraints are applied are the basis of the *control strategy* for the construction of token aggregations. Tokens of different types can be aggregated in many ways and one seeks to avoid the combinatorics of computing relations between large token sets. There are two obvious ways in which the constraint functions can be used in control. A constraint on the cross product of the token sets can be used either to *rank order* the set of token

tuples via the response of the constraint function, or to *filter* (i.e. select) a subset of the token tuples for further processing. Of course the responses of the constraint functions that are used for ordering could also be used for filtering by specifying filtering criteria such as thresholds. Example control strategies for limiting the computation are outlined in [RIS87].

5 Results

This section presents the results of forming aggregations via token attribute and token relational constraints applied to suburban house scenes and road scenes; Figure 5 shows a representative image of one of the house scenes and a typical region segmentation and line description for the image. The examples chosen for this paper involve aggregations of tokens that serve as texture measures and aggregations of tokens with specific shape properties.

A simple texture measure based on line density can be computed by counting the number of lines within a region and normalizing by the region size. This is accomplished by forming

a

b

c

a

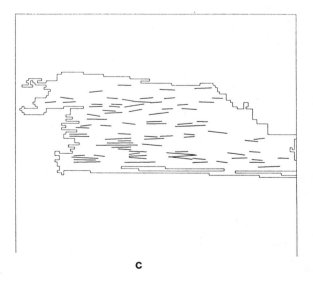

b

c

Figure 5. Results. (a) a black and white rendering of the original color image; (b) regions produced by a segmentation system using localized histograms followed by region merging; (c) straight lines produced by an algorithm which uses similarity of gradient orientation as the primary organizational feature.

aggregations of regions and their interior lines. A filtering constraint uses the interior-line-percentage relational measure to select only those lines which are completely (or mostly) interior to the region. A filter is defined to group into an aggregation those lines associated with each region which sufficiently satisfy both relational and attribute constraints. The density of interior lines in each aggregation is then computed as an attribute of this new token and mapped to a score for the region (which can also be thought of as a score for the region-line aggregation).

Figure 6a shows an example of extracting interior lines for regions in a house scene, and then computing the interior line density of these regions as a texture measure (see Figure 6b). Some objects, notably the roof of a house, are characterized by short horizontal lines (due to the shingles) interior to the region. By adding the additional attribute constraint of horizontal orientation on the interior lines, the previous result can be extended to focus attention on the house roof as shown in Figure 6c,d. Additional constraints on line length and line contrast can be defined to extract only short, horizontal interior lines to the degree that these characteristics of the expected texture element are known.

The roof region could be obtained or verified in another way. The line-boundary-percentage relational measure could be used to select lines which lie to a great extent on the boundary of the region (see Figure 7a). A line attribute rule could then be defined to favor long lines (see Figure 7b). The lines which received high scores from both the line boundary rule and the line length rule (i.e., long boundary lines) could then be grouped to form a region-boundary aggregation. At that point parallel relations, rectangle or parallelogram structures could be identified.

Another simple shape measure can be computed by determining if a region is bounded by a pair of long vertical lines. As Figure 8 shows, the process is useful for extracting telephone poles in road scenes. The filtering relational constraint for this measure uses line-boundary-percentage to select only those lines which lie on a region boundary; an attribute constraint then selects the long vertical lines. Relational measures can be defined

d

Figure 6. Example of Texture Measure for Extracting House Roof. A simple texture measure computed by a relational constraint based upon the density of lines within a region. (a) Lines which received a high score on the relational measure of interior-line-percentage (i.e. INTERIOR lines); (b) The density of interior lines for each region represented by the density of shading; (c) Horizontal INTERIOR lines for the roof region; (d) Density of horizontal interior lines.

to form aggregations of pairs of parallel overlapping lines from the long vertical boundary lines.

A variety of more complex 2D shapes can be matched to lines by extracting lines that bound regions. The techniques presented here are only a part of more complex grouping and model matching procedures that are being developed in other research [HAN87a,REY87].

Figure 9 shows a set of rectangles extracted from a house scene. In this case, the set of lines intersecting a region were filtered to extract the set of bounding lines. These were further filtered on the basis of co-parallel, collinear, and endpoint coincidence relations. Pairs of 'adjacent' lines were then filtered on the basis of constraints on their relative orientation in order to form corner hypotheses and the resulting set of lines were matched to a rectangle model. As the figure indicates partial matches to the rectangle model are allowed. The matches could be further restricted if one is seeking dark shutters by using a constraint on the intensity of regions, although this was not done in this example.

There are sometimes serious problems with constructing aggregations through relational measure directly computed from initial token representations. If the desired primary token is fragmented, whether it be a region or a line, then the expected relational responses might be distorted significantly because some of the expected token attributes, token relations, and features of the extracted aggregation may be significantly changed. One must balance the unreliability of extracting useful primary tokens by the computational savings achieved by focussing upon the subset of secondary tokens that satisfies some relational constraints with respect to the filtered primary set. To the degree that these problems occur, many stages of hierarchical token aggregation may be necessary, perhaps using more complex strategies for applying relational measures and grouping tokens.

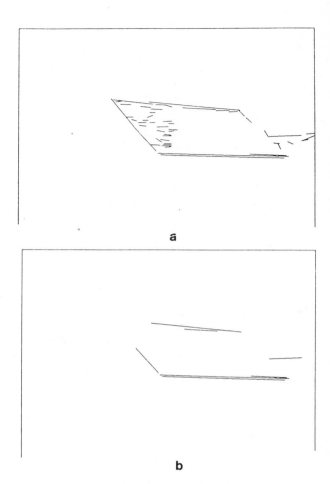

a

b

Figure 7. Extracting Roof via Long Bounding Lines. Given a possible roof region the long bounding lines can be filtered to find a roof shape. (a) Lines bounding the hypothesized roof region, and (b) Long bounding lines which can be the basis of forming an approximate parallelogram shape.

Let us consider a specific real example, shown in Figure 10, where both representations (e.g. regions and lines) would have some difficulty in directly providing the basis for aggregations of tokens from the other representations. The roof region in Figure 5b is fragmented into many smaller regions (Figure 10a). In this case no region will get full benefit of the bounding roof lines, since each region is only bordered by a subset of the roof lines; Figure 10b depicts the lines intersecting the largest roof region. An alternate grouping strategy would be to extract a set of roof lines first (which may be a difficult task in itself), then the set of regions that are bounded by these lines could be used in some manner to aggregate regions. The line information if properly filtered shows the outline of the roof fairly clearly (Figure 10c), but the initial set of lines in that spatial vicinity provide many possible aggregations. The process of grouping lines into meaningful geometric structures is a non-trivial problem and is a focus of continuing work on grouping and knowledge-directed processing in our research group [BOL87,HAN87a,HAN87b,REY87,WEI86,WEY86].

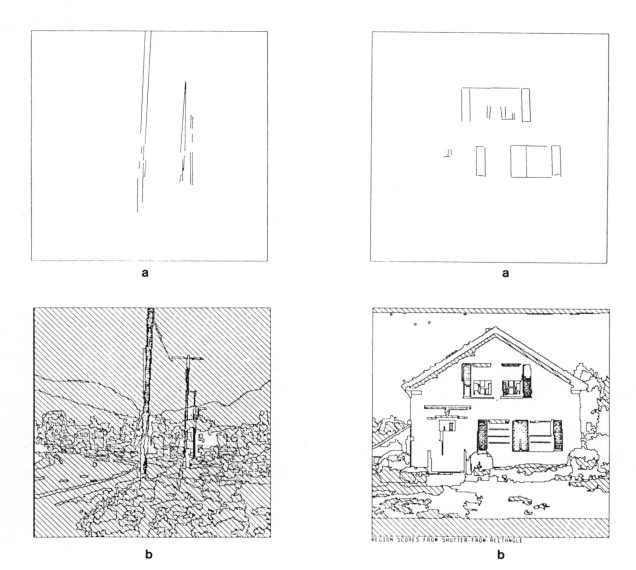

a

b

a

REGION SCORES FROM SHUTTER-FROM-RECTANGLE

b

Figure 8. Example of Extracting Telephone Poles via Vertical Bounding Lines. (a) The line pairs formed by the line constraint of bounding vertical lines. (b) The relational measures are mapped back to the regions; hatched regions have no long vertical bounding lines and are vetoed.

Figure 9. Extracting Rectangular Window and Shutter Hypotheses. (a) Horizontal and vertical bounding lines; (b) The regions produced by a relational constraint on the extracted bounding lines.

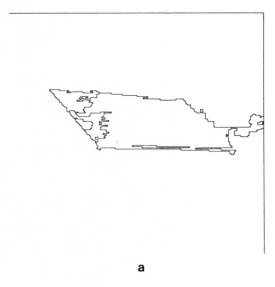

a

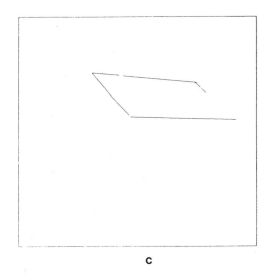

c

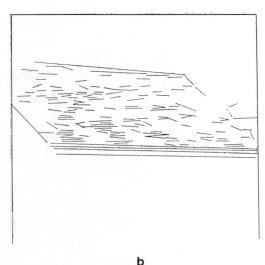

b

Figure 10. Example of difficulties in aggregating tokens. (a) The roof is fragmented in the region representation; (b) The lines intersecting the largest roof region only capture a portion of the relevant line information; note the two key missing lines at the upper left corner of the roof. (c) A subset of the full set of lines which, if they could somehow be selected, would provide the appropriate line aggregation to group the roof regions, and allow the roof outline to be completed in a straightforward manner [WEY86].

6   Conclusion

The use of constraints on relational measures between tokens of the same and different types is a uniform, straightforward way of combining information from multiple low-level processes. The techniques developed in this paper allow information fusion to take place during the interpretation process as intermediate tokens are aggregated via object-dependent constraints on token attributes and token relations. The ideas presented here can be used to build hierarchical aggregations; for example, aggregates of lines could be formed by grouping colinear sets of lines into new longer line tokens [e.g.,BOL87,REY87b,WEI86]. By treating each aggregation as a new token, attributes could then be computed for each and the constraints applied recursively.

The types of information that could be added include surface segmentations, and 2D and 3D motion and depth token attributes. Each segmentation or low-level process would create a set of tokens with associated attributes which could be added to the intermediate-level representation. These new tokens could then be used in the same way as regions and lines are used now. Each new token type would require the definition of relational measures between tokens of different types. There would be no other major modifications to the system. The approach overlaps issues and techniques in the areas of grouping and model matching. To the degree that tokens in one of the several representations do not exhibit the characteristics that provide the basis for directly extracting the desired structures, more complex perceptual organizing processes and knowledge-based strategies will be required. Each object that must be recognized could be defined by a separate model and control strategy for aggregating the different token types. While the concepts of relational measures and relational constraints can still be the basis of these strategies, many stages of hierarchical aggregations may be required. In such cases efficient control strategies will become a major issue. Thus, multiple alternate grouping strategies will probably be required in order to extract and utilize information across multiple representations in a generally robust and efficient manner. In the VISIONS system [DRA87a,b,HAN87a], the knowledge-based schema system provides flexible mechanisms for defining and applying control strategies. The mechanisms described in this paper are only meant to serve as the first stage of this organizing process.

7 Acknowledgements

The authors would like to express their appreciation to Amerinex Artificial Intelligence Corporation (AAI) for KBVision, a commercially available, integrated, graphically-oriented environment supporting basic research at all three levels of the image understanding hierarchy and to Rob Belknap (AAI) for many of the results presented in the paper. They would also like to thank Laurie Waskiewicz at the University of Massachusets for her efforts in preparing this manuscript.

References

[ARK87]   R. Arkin, "Towards Cosmopolitan Robots: Intelligent Navigation in extended Man-Made Environments", Ph.D. dissertation, Computer and Information Science Department, University of Massachusetts at Amherst, expected in August 1987.

[BEL86]   R. L. Belknap, E. M. Riseman, and A. R. Hanson, "The Information Fusion Problem and Rule-Based Hypotheses Applied to Complex Aggregates of Image Events," *Proc. IEEE Conference on Computer Vision and Pattern Recognition*, Miami, Fl., June 1986.

[BES85]   P.J. Besl, and R.C. Jain, "Range Image Understanding", in *Proc. IEEE Conference on Computer Vision and Pattern Recognition*, San Francisco, CA, June 9-13, 1985, New York, pp. 430-451.

[BEV87]   R. Beveridge, A. Hanson, and E. Riseman, "Segmenting Images Using Localized Histograms and Region Merging", submitted to *International Journal on Computer Vision*, Spring/Summer 1988, also COINS Technical Report 87-88, Computer & Information Science Department, University of Massachusetts at Amherst.

[BOL87]   M. Boldt and R. Weiss, "Token-Based Image Abstraction", COINS Technical Report, Computer & Information Science Department, University of Massachusetts at Amherst, July 1987.

[BUR86]   J. B. Burns, A. R. Hanson, and E. M. Riseman, "Extracting Straight Lines," *IEEE Transactions on Pattern Analysis and Machine Intelligence*, vol. PAMI-8, pp. 425-455, July 1986.

[DRA87a]   B.A. Draper, R.T. Collins, J. Brolio, J. Griffith, A. Hanson, and E. Riseman, "Tools and Experiments in the Knowledge-Directed Interpretation of Road Scenes", *Proc. of the DARPA Image Understanding Workshop*, Los Angeles, CA, February 1987, pp. 178-193. Also COINS Technical Report 87-05, Computer & Information Science Department, University of Massachusetts at Amherst, January 1987.

[DRA87b]   B.A. Draper, J. Brolio, R. Collins, A. Hanson, and E. Riseman, "The Schema System", submitted to *International Journal on Computer Vision*, Spring/Summer 1988, also COINS Technical Report, Computer & Information Science Department, University of Massachusetts at Amherst, in preparation.

[HAN78a]   A. R. Hanson and E. M. Riseman, "VISIONS: A Computer System for Interpreting Scenes," *Computer Vision Systems* (A. Hanson and E. Riseman, eds.) (1978), pp. 303-333, Academic Press.

[HAN78b]   A. R. Hanson and E. M. Riseman (Eds.), *Computer Vision Systems*, New York, Academic Press, 1978.

[HAN87a]   A. R. Hanson and E. M. Riseman, "The VISIONS Image Understanding System - 1986", in *Advances in Computer Vision*, (Chris Brown, Ed.), Erlbaum Press, also COINS Technical Report 86-62, Computer & Information Science Department, University of Massachusetts at Amherst, December 1986.

[HAN87b]   A.R. Hanson and E.M. Riseman, "From Image Measurements to Object Hypotheses", COINS Technical Report 87-129, University of Massachusetts at Amherst.

[KIT86]   L. Kitchen, R. Weiss and J. Tuttle, "Identification of Human Faces Using Data-Driven Segmentation, Rule-Based hypothesis Formation, and Interative Model-Based Hypothesis Verification", COINS Technical Report 86-53, Computer & Information Science Department, University of Massachusetts at Amherst, October 1986.

[KOH83]   R. R. Kohler. "Integrating Non-Semantic Knowledge into Image Segmentation Processes," Ph.D. Thesis, University of Massachusetts at Amherst, September 1983; also COINS Technical Report 84-04.

[MCK84]   D. M. McKeown, W. A. Harvey and J. McDermott, "Rule Based Interpretation of Aerial Imagery," Dept. of Computer Science, Carnegie-Mellon University, (September 1984).

[MCK85]   D. M. McKeown, W.A. Harvey and J. McDermott, "Rule Based Interpretation of Aerial Imagery", *IEEE PAMI*, Vol. PAMI-7, No. 5, September 1985, pp. 570-585.

[MCK86]   D. M. McKeown, C.A. McVay and B.D. Lucas, "Stereo Verification in Aerial Image Analysis", *Optical Engineering*, Vol. 25, No. 3, March 1986, pp. 333-346, Also available as Technical Report CMU-CS-85-139.

[NAG82]   P. A. Nagin, A. R. Hanson, and E. M. Riseman, "Studies in Global and Local Histogram-Guided Relaxation Algorithms," *IEEE Transactions on Pattern Analysis and Machine Intelligence* **3** (May 1982), pp. 263-277.

[NAN87]   N. Nandhakumar and J.K. Aggarwal, "Multisensor Integration–Experiments in Integrating Thermal and Visual Sensors", *Proc. IEEE ICCV*, London, England, June 8-11, 1987, pp. 83-92.

[REY87a]   G. Reynolds, N. Lehrer and J. Griffith, "A Method for Initial Hypothesis Formation in Image Understanding", *Proc. of the DARPA Image Understanding Workshop*, Los Angeles, CA, February 1987. Also COINS Technical Report 87-04, Computer & Information Science Department, University of Massachusetts at Amherst, January 1987.

[REY87b]   G. Reynolds and J. Ross Beveridge "Geometric Line Organization Using Spatial Relations and A Connected Components Algorithm", COINS Technical Report, Computer & Information Science Department, University of Massachusetts at Amherst, in preparation, 1987.

[RIS84]   E. M. Riseman and A. R. Hanson, "A Methodology for the Development of General Knowledge-Based Vision Systems," *IEEE Proc. of the Workshop on Computer Vision: Representation and Control* (1984), pp. 159–170.

[RIS87]   E. M. Riseman and A. R. Hanson, "A Methodology for the Development of General Knowledge-Based Vision Systems", in *Visions, Brain, and Cooperative Computation*, (M. Arbib and A. Hanson, Eds), MIT Press, 1987, pp. 285-328. Also COINS Technical Report 86-27, University of Massachusetts at Amherst, July 1986.

[SHA86]   S.A. Shafer, A. Stentz, and C.E. Thorpe, "An Architecture for Sensor Fusion in a Mobile Robot", *International Conference on Robotics and Automation*, San Francisco, CA, 1986, pp. 2202-2011.

[WEI86]   R. Weiss and M. Boldt, "Geometric Grouping Applied to Straight Lines", *Proc. of the IEEE Computer Society Conference on Computer Vision and Pattern Recognition*, Miami, FL, June 22-26, 1986, pp. 489-495.

[WEY83]   T. Weymouth, J. Griffith, A. R. Hanson, and E. M. Riseman, "Rule Based Strategies for Image Interpretation," *Proc. AAAI-83* (August 1983).

[WEY86]   T. E. Weymouth, "Using Object Descriptions in a Schema Network for Machine Vision," Ph.D. Thesis, University of Massachusetts at Amherst (April 1986). Also COINS Technical Report 86-24, Computer & Information Science Department, University of Massachusetts at Amherst, May 1986.

[WIL81]   T. Williams, "Computer Interpretation of a Dynamic Image from a Moving Vehicle," Ph.D. Thesis and COINS Technical Report 81-22, University of Massachusetts at Amherst (May 1981).

Autonomous reconfiguration of sensor systems using neural nets

Oleg G. Jakubowicz

State University of New York, Electrical and Computer Engineering Department

ABSTRACT

Neural networks are ideally suited for sensing images and waveforms, processing them into intermediate levels of representation and outputing identification and/or characteristics of the sensed object. These networks can solve problems that conventional algorithms haven't and already in several cases this new technology has performed better than humans (e.g. sonar signal classification).

A brief review of where autonomous agents may use neural networks and their learning algorithms is presented. A high yielding area is seen in the self-repair of damaged or faulted components. Architectures are proposed for implementing self-repairing sensor and identification systems aboard autonomous agents.

One example is presented for a system which identifies visual objects. This system has four layers of massively connected simple parallel processors. Each connection has a weight attribute and the collected assignment of weights in a layer determines what function the layer will perform.

The first layer (the imput layer) is simply the pixel detector layer. The second layer has eight sublayers which are sensitive to short line segments in eight different orientations. The third layer detects elementary combinations of the lower lines such as oriented corners or curve segments. The fourth layer has one sublayer for each macroscopic object to be identified which may be fused with a pinpoint location sensor.

The crux of using reconfiguration in this type of sensor is that when one (or several) of the units or detectors become inoperative then neighboring detectors in that layer may be used to reprogram the weights connecting surviving units to restore functionality. This strategy takes advantage of the redundancy of parallel processors present in most types of neural networks. Alternatively a properly functioning agent may teach the injured agent or competitive learning for repairing middle processing layers may be utilized when an operative after-the-fact sensor is available for teaching the output layer.

1. INTRODUCTION

1.1 Autonomous agents' need can be fulfilled by neural nets

Autonomous agents (a.a.'s) are intelligent robots which operate in isolated locations. These a.a.'s function is to recognized states of their environments, classify situations and objects and to perform some action in response to it. Examples of isolated environments are robots on other planets, satellites or unmanned submarines which need to keep their presence hidden or deep mining probes. The common need for a.a.'s to remain functionable when a simple component is disabled or incorrectly used or an unforeseen situation develops can be met excellently by the basic ability of neural networks to perform graceful degradation and generalization.

1.2 Elementary example of an autonomous agent

An example of a large neural network with several detectors or sensors (oriented line detectors), a degree of redundancy and hierarchical levels of processing is Fukushima's Neocognitron[1,2] and the author's VLIV vision system[3]. In VLIV there are eight line detectors/sensors (one oriented at every 22.5 degrees) each of which feeds forward into each of twelve second layer detectors for elementary combinations of the lines. These feed into a third layer of units which incorporate lateral inhibition for object of middle complexity. A top layer is able to recognize relationally described objects. This could be used as a prototypical example of an a.a. where a different output action is called for by each type of object recognized.

1.3 Some lessons learned from neural networks

Neural networks have been built that perform both as well as and better than human beings. An example is provided by Gorman and Sejnowski[4] in the classification of sonar signals. By using nonlinear learning with a teacher method they were able to classify the training set of the frequency spectrums of sonar signal returns from rocks and metal spheres 99% correctly as opposed to human sonar technicians who classified 88 to 93% correct. The network was 92% on a subsequent testing set.

Neural networks have the capability to generalize but occassionally they generalize to form an improper response. This is well described by a recent neural network developed by Tesauro and Sejnowski[5] which learned to play backgammon from a human expert. Though the network would learn to make good moves in most situations, it occassionally made silly blunders. An autonomous agent can not afford blunders but a supervisory rule based system should monitor the output from the neural need and supersede unqualifying actions.

1.4 Neural nets have desirable properties for autonomous agents

There are two basic ways a neural net can be inflicted and damaged. The hardware might be fully operational and only lost its memory, neural net weights may be erased. Secondly neural net nodes may be destroyed, some hardware is permanently disabled. Neural nets are known to degrade gracefully with such damage and erasure. Furthermore reprogramming alias relearning is still possible in such systems. Furthermore the relearning will require only a small fraction of the time or iterations of the original learning. Numerous examples are given in McClelland and Rumelhart's books [6].

2. NEURAL NETWORKS IN AUTONOMOUS AGENTS

2.1 Autonomous agent types and functions

Autonomous agents can be either open ended or closed ended. The closed ended type is simpler in that it will respond only to predetermined situations. The output or response can be a number of fixed actions. The opened ended type of a.a. has a higher order of processing in that it needs to either recognize an heretofore unrecognized situation or react and respond to an input situation that was not previously programmed. An example of a type 1 a.a. is a telephone switchboard router or an operating system of a computer. An example of a type II a.a. could be a robot on another planet that could not wait for a reply from earth in a perilous unforeseen situation such as starting to fall off a cliff. Another example of a type II a.a. is a robotic submarine or satillite which senses an unusual situation of the dangerous kind and can not communicate for help or interpretation for fear of being disclosed.

Type II autonomous agents' programmed responses are usually stated as goals to be fulfilled. These may be accomplished by appropriately performing one or several actions chosen from a limited number of actions. The actions are usually composed of a limited number of subactions, functions or steps. Therefor often the implicit goal of an a.a. is to produce a plan with a high likelihood of successfully satisfying its explicit goals. Therefore a evaluation of probable success criteria function or method is needed, which in current technology would mostlikely be a rule-based expert system. The expert system would have the evaluation reasoning of the robot's owner. A general layout of both a type I and a type II a.a. utilizing neural nets is presented in figure 1. The large dark block contains the additional necessary modules for a type II a.a.

2.2 General places for neural nets in autonomous agents

Neural networks would serve a.a.'s most profitably in:
1. the middle levels of signal and image processing where weight (monotonically proportional to probability of likelyhood) needs to be given to various alternatives and/or a 'strongest' alternative must be picked (winner-take-all layer),
2. the middle levels of response decision making where once again weight of alternatives and/or 'strongest' alternatives must be decided upon.

In the signal and image processing layers neural nets are finding applications such as local feature sensors, integrative feature sensors, segmentation mappers, elementary object detectors and modelled object detectors. Their inputs may come from several similar detectors or sensors.

In the integrative levels neural nets may be used to make a decision or classification based on the inputs of different kinds of sensors such as stereo sensor/detectors, motion detectors, shape from optical vision and shape from synthetic aperature radar receivers. A neural net can also be used in the generation of an response action or plan and will probably operate under the auspices of a knowledge based directing system to assure uncanny responses are not generated.

At first the early processing might be done with conventional signal processing and only the detecting of intermediate and high level features and objects along with the integration of information from different sensor types may be implemented in an a.a.. But later the early signal processing and even the high level rule based processing components may be implemented as neural nets to take advantage of graceful degradation, generalization and relearning capabilities of neural networks.

In a type II a.a. new plans or actions (that are in accord to the top level goals of the syste) may be constructed. This will probably be done by a knowledge based heuristic system not much unlike Lenat's AM and Erisko programs [7], Holland's genetic algorithms [8] or other concept learning systems . Associative neural nets

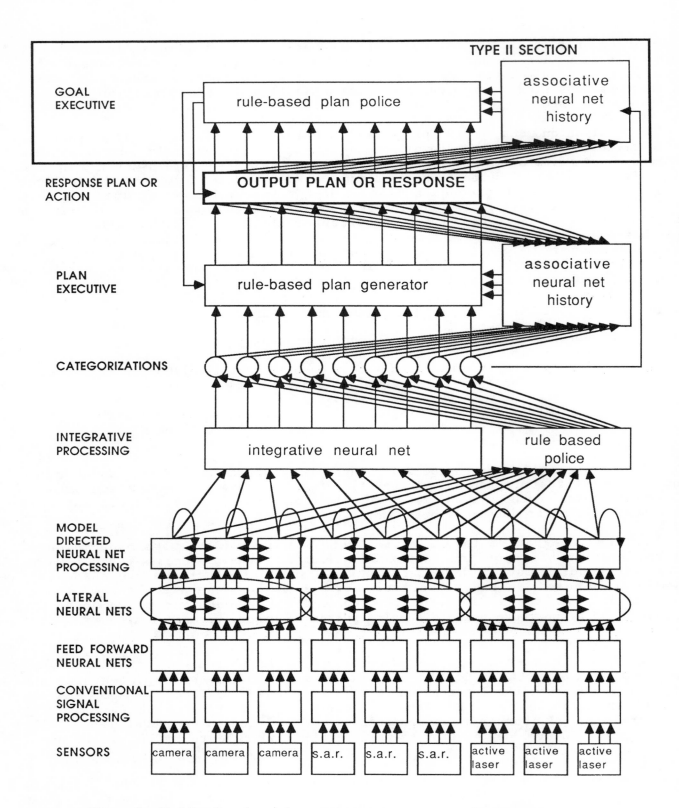

Figure 1. Architecture for autonomous agents with neural net components.

can be trained to produce the same response to a given set of categorizations at the integrative neural net. However due to the fact that the neural net might not have learned every rule or fact in the rule based goal overseer and the advantageous characteristic of generalization the neural net could serve the useful purpose of providing suggestions for the rule based system's search procedure for a goals satisfying response/plan.

2.3 Repair requires relearning - learning methods

Learning schemes can usually be classified into one of three general methods; competitive, backward error propagation and stocastic, or single instance learning. Competitive learning is an unsupervised technique in which the outputs classification units initially respond randomly but through iterations each output unit tends to respond to the input pattern closest to initial random pattern. Besides requiring many iterations the output units are not prespecified and will not be correlated with a next higher layer of neural-like processing unless output units are somehow identified with inputs. Then either the output units need to be reorganized to 'fit' into the next layer's processing input expectations or the next layer's input must be rearranged.

Backward error propagation and stocastic learning methods usually provide excellently learned neural nets but they require enormous amounts of processing of the required iterations. The input and output training set must be recorded and repeatedly presented to the system in order to learn. In a damaged or erased a.a. the hugh memory required to store the training instances is most likely to be just as damaged as the neural net itself. Therefore the training instances must be provided by another a.a. or repairing unit. Neither should it be possible for him to learn on his own by observation (with other of his components providing the teacher function) since this would require such a large set of observations and time.

Several models have been proposed where a reserved bank of nodes or units is available for learning subsequent training examples. Often these models will use just one instance of a training prototype to teach the new unit. Often these models are taught by these single instances by human preprogramming (see Fukushima [2], Nexus [10] and Jakubowicz [3]). An erased a.a. can preprogram each different type of output unit via single instance presentation, either by a recorded example or one provided by the environment and a teacher.

3. SYSTEMS FOR SELF REPAIR

Two systems are presented that can allow self repair of an a.a.'s in an isolated environment.

3.1 Neighboring detectors and processors providing the exemplar and teacher

Often there is a great deal of redundancy especially in the lower levels in processing. There may be duplicate detector systems whose only purpose is to aid in the detection faint signals or be explicitly available for repairing. If the weight s of one neural net are erase then a detector that can classify the same information can be used to provide the teacher for observed exemplars. Mutually if the second neural net developed erase weights the first could train the second. All weights in the neural net could be kept with only the erased ones allowed to change to improve the network's performance. If some of the nodes or inputs were destroyed then as illustrated above the weights often just need to be slightly shifted by a few iterations of backward error propagation training. If the system was taught by a single instance method then just one instance of the feature(s) corresponding to the damaged processor(s) will be sufficient for reprogramming.

3.2 Delayed training by the autonomous agent's processors and detectors

An a.a. may have a detector that can identify an input state or pattern after the fact. For example a robot might not know he's falling until after he has fallen or a a.a. submarine may not recognize an adversive head scout until he's seen the party that the scout is leading. If a neural net that could identify the input in real time is damaged then he could use the late detector's classification as his teacher. His teacher would not be able to teach him until after the fact but after he's completed his training the restored detector will be able to recognize the (maybe perilous) situation before its too late for the a.a. to do something about it.

4. HOSPITAL SYSTEMS

If redundancy is not buildt into the a.a. and there is more than one a.a. isolated from the rest of the world but not from each other then one functioning a.a. can come to the aid of the damaged a.a. to become his temporary teacher. A helpful a.a. with equivalent detectors and processors could aid in the repair of the damaged a.a. functioning as a redundant neural net as described above without needing to translate information from one detector to the other. If the trainer and trainee are not of the same type but do detect similar instances then there would be some translation of input and teacher vectors involved also. An aiding a.a. with slower or less precise detectors could still be of help by methods similar to the delayed teacher method above. If these a.a.'s were able to help each other in their isolated environment then they would have a much better chance of surviving as a functioning autonomous agent.

5. REFERENCES

1. Fukushima,K., Miyake,S. and Ito,T., Neocognitron: A Neural Network Model for a Mechanism of Visual Pattern Recognition, IEEE Trans. Systems, Man, Cybernetics, 5, p.826-834, 1983.
2. Miyake,S. and Fukushima,K., A Neural Network Model for the Mechanism of Feature Extraction, Biological Cybernetics, 50, p.377-384, 1984.
3. Jakubowicz,O., A Connectionist Structured Object Recognizer for Noisy Images, in the Proceedings of Applications of Artificial Intelligence VI, 1988.
4. Gorman,P. and Sejnowski,T., Learning Classification of Sonar Targets Using a Massively Parallel Network, Workshop on Neural Network Devices and Applications, Jet Propulsion Laboratory, Pasadena, CA, Feb. 18-19, 1987.
5. Tesauro,G. Sejnowski,T., A 'Neural' Network That Learns to Play Backgammon, IEEE Conference on "Neural Information Processing Systems - Natural and Synthetic", Denver, CO, Nov. 8-12, 1987.
6. McClelland,J. and Rumelhart,D., Parallel Distributed Processing Vol. I and II, MIT Press, 1986.
7. Lenat,D., Eurisko: A Program That Learns New Heuristics and Domain Concepts, Artificial Intelligence, 21, p.31-98, 1983.
8. Holland,J., Holyoak,K., Nisbett,R. and Thagard,P., Induction: Processes of Inference, Learning, and Discovery, MIT Press, 1986.
9. Machine Learning Vols. I and II, edited by Michalski,R., Carbonell,J. and Mitchell,T., Morgan Kaufmann Publishers, Inc., 1983 and 1986.

10. Bradshaw,G., Learning About Speech Sounds: The NEXUS Project, Proceedings of the Fourth International Workshop on Machine Learning, University of California, Irvine, CA, p. 1-11, June 22-25, 1987.

Multisensor Knowledge Systems

Thomas C. Henderson

Department of Computer Science
University of Utah
Salt Lake City, Utah 84112

ABSTRACT

We describe an approach which facilitates and makes explicit the organization of the knowledge necessary to map multisensor system requirements onto an appropriate assembly of algorithms, processors, sensors, and actuators. We have previously introduced the Multisensor Kernel System and Logical Sensor Specifications as a means for high-level specification of multisensor systems. The main goals of such a characterization are: to develop a coherent treatment of multisensor information, to allow system reconfiguration for both fault tolerance and dynamic response to environmental conditions, and to permit the explicit description of control.

In this paper we show how Logical Sensors can be incorporated into an object-based approach to the organization of multisensor systems. In particular, we discuss:

* a multisensor knowledge base,

* a sensor specification scheme, and

* a multisensor simulation environment.

1. INTRODUCTION

The rapid design of embedded electromechanical systems is crucial to success in manufacturing and defense applications. In order to achieve such a goal, it is necessary to develop design environments for the specification, simulation, construction and validation of multisensor systems. Designing and prototyping such complex systems involves integrating mechanical parts, software, electronic hardware, sensors and actuators. Design of each of these kinds of components requires appropriate insight and knowledge. This in turn has given rise to special computer-based design tools in each of these domains. Such Computer Aided Design (CAD) systems have greatly amplified the power and range of the human designer. To date, however, it is still extremely difficult to address overall system issues concerning how the components fit together, and how the complete system will perform.

It is crucial to develop a design environment in which these multiple facets of system design can take place in a coordinated way such that the description of one component can be easily interfaced to another component, even when they are radically different kinds of things (e.g., a control algorithm, a mechanical linkage and an actuator). The designer should have the freedom to try out ideas at different levels of detail; i.e., from the level of a sketch to a fully detailed design. The Multisensor Knowledge System provides part of the solution to developing such an environment.

Logical Sensor Specifications (LSS) were developed previously as a method to permit an implementation independent description of the required sensors and algorithms in a multisensor system. The basic unit is the *logical sensor*. Sensor data flows up through the currently executing program (one of $program_1$ to $program_n$) whose output is characterized by the *characteristic output vector*. Control commands are accepted by the *control command interpreter* which then issues the appropriate control commands to the Logical Sensors currently providing input to the selected program. The programs 1 through n provide alternative ways of producing the same characteristic output vector for the logical sensor. The role of the *selector* is to monitor the data produced by the currently selected program and the control commands. If failure of the program or a lower level input logical sensor is detected, the selector must undertake the appropriate error recovery mechanism and choose an alternative method (if possible) to produce the characteristic output vector. In addition, the selector must determine if the control commands require the execution of a different program to compute the characteristic output vector (i.e., whether dynamic reconfiguration is necessary).

Logical Sensor Specifications are useful then for any system composed of several sensors, where sensor reconfiguration is required, or where sensors must be actively controlled. The principle motivations for Logical Sensor Specifications are the emergence of significant multisensor and dynamically controlled systems, the benefits of data abstraction, and the availability of smart sensors.

In previous papers we have explored several issues of multisensor integration in the context of Logical Sensor Specifications:

* fault tolerance [8],

* functional (or applicative) style programming [16],

* features and their propagation through a network [17],

* the specification of distributed sensing and control [9, 10],

* the automatic synthesis of Logical Sensor Specifications for CAD/CAM applications [11, 12].

Related work includes that of Albus [1] on hierarchical control, Bajcsy et al. [2] on the Graphical Image Processing Language, Overton [15] on schemas, and Chiu [4] on functional language and multiprocessor implementations. For an overview of multisensor integration, see Mitiche and Aggarwal [13].

In exploring these issues, we have found that the specification of multisensor systems involves more than just sensor features. It is true that knowledge must be available concerning sensors, but it is essential to also be able to describe algorithms which use the sensor data and the hardware on which they are executed. In the rest of the paper, we describe the components of an object-based approach to developing a knowledge system to support these requirements.

2. OBJECTS AND METHODS

Several distinct programming styles have been developed over the last few years, including:

* **applicative-style programming,**

* **control-based programming,**

* **logic programming,** and

* **object-based programming.**

Applicative style programming exploits function application as its main operation and regulates quite strongly the use of side-effects [6]. Historically, however, control-based programming has been the most extensively used paradigm, and focuses on the flow of control in a program. Logic programming is based on logical inference and requires the definition of the formal relations and objects which occur in a problem and the assertion of what relations are true in the solution. On the other hand, many current systems are being developed which are based on the notion of objects; this style emphasizes data abstraction combined with message passing [3, 14].

In the control-based style a program is viewed as a controlled sequence of actions on its total set of data structures. As the complexity of a system grows, it is hard to keep a clear picture of the entire sequence of actions that make up the program. This leads to the chunking of sequences into subprograms, and this is almost exclusively done for control purposes. But data structures are not decomposed into independent entities. In fact, most global data structures are shared by all subroutines.

On the other hand, the object-based style takes the view that the major concern of programming is essentially the definition, creation, manipulation and interaction of objects; that is, a set of independent and well-defined data structures. In particular, a single data structure (or instance) is associated with a fixed set of subprograms (methods), and those subprograms are the only operations defined on that object.

Such a use of data abstraction leads to design simplification which in turn makes the program more understandable, correct, and reliable. In addition, flexibility and portability are enhanced since details of objects (i.e., their representations) are hidden and can be implemented in other ways without changing the external behavior of the object.

For our purposes, an object consists, essentially, of three parts:

1. <u>unique</u> <u>name</u>: this name must be distinguished from all other names in both time and space,

2. <u>type</u>: an object is an instance of a type which defines the valid set of operations and which details the nature of the resource represented, and

3. <u>representation</u>: the representation contains the information content associated with an object. This may include private data structures, references to other objects, etc.

Thus, an object is a structure with internal state (perhaps called *slots* and comprised of name/value relationships) accessed through functions (also called *methods*) defined in association with the object. This approach makes management schemes simpler and fewer, easier to implement and use; in addition, individual resources are easier to specify, create (allocate), destroy (deallocate), manipulate and protect from misuse.

It has been effectively argued many times that object-based programming is well-suited to embedded systems processing requirements. In particular, the application of this methodology to the specification of sensor systems helps to directly describe most of the important aspects of such systems:

* parallel processing,

* real-time control,

* exception handling, and

* unique I/O control.

Sensors typically require such operations as: enabling/disabling, limit setting, status checking, and periodic logging of state. That is, sensor systems must respond to out-of-limit readings and issue alarms, detect faulty sensors, and recover from failure, and these functions can be implemented in a straightforward way.

3. MULTISENSOR KNOWLEDGE SYSTEMS

Much of our previous work on multisensor systems has concentrated on the specification of such systems and reasoning about their properties. It is necessary to be able to describe both the parameters and characteristics of individual components of multisensor systems, and to be able to deduce global properties of complete systems. Although it may be possible to deduce such properties (especially static properties like complexity, data type coercion, etc.), we believe that many interesting properties can only be determined by simulating the operation of the complete system.

Thus, we seek a representation that supports:

1. **multisensor system specification**: this describes the components and interconnection scheme of the particular system being designed,

2. **sensor, algorithm, processor and actuator knowledge representation**: this structures information about sensor characteristics (e.g., accuracy, hysteresis, dynamic range, etc.), algorithms (e.g., space and time complexity, amenity to parallel computation, stability, etc.) processors (e.g., cycle times, memory limits, address space, power requirements, etc.), and actuators (e.g., actuation principle, power requirements, etc.), and

3. **multisensor system simulation**: this permits one to monitor important parameters and to evaluate system performance.

In the following subsections, we describe the Multisensor Knowledge System (**MKS**), an object-based approach to providing a unified answer to these three capabilities.

3.1 THE MULTISENSOR KNOWLEDGE BASE

The multisensor knowledge base serves two main purposes:

1. to describe the properties of the system components (e.g., sensors, algorithms, actuators and processors), and

2. to provide class descriptions for the actual devices which are interconnected in any particular logical sensor specification.

That is, the knowledge base must describe not only generic sensors (e.g., cameras), but specific sensors (e.g., Fairchild 9000, Serial No. 28753). It is then possible to reason about sensor systems at several levels. Moreover, it is possible that two distinct specifications require some of the same physical sensors. In such a case, it is the responsibility of the execution environment to resolve resource allocation conflicts.

We have chosen a frame-like knowledge representation. Frames relate very naturally to object-based descriptions, and, in fact, can be viewed as a class of restricted objects. It is straightforward to provide hierarchical descriptions of system components. For example, a *CCD Camera* frame has two slots: element spacing and aspect ratio. These slots are specific to CCD cameras and as such do not appear as slots for *2-D cameras*. These latter have slots for scanning format, scan timing, resolution, output signal, and operating conditions. These slots are inherited by any instance of CCD camera. One level up, we find a frame for *Vision* sensors. This frame has specific slots for the spectral band and for the output type (e.g., 2-D byte array, multi-band, etc.). At the highest level of the hierarchy is the *Sensor* frame which has a slot for the physics of operation. This slot is used by any particular sensor to allow for an explanation of the physics behind the workings of the sensor. In this way, if reasoning is required about the sensor, it is possible to look in this slot for information. As can be seen, knowledge is organized such that there are more specific details lower in the hierarchy.

Note that frames are themselves implemented as objects. Thus, actual devices are instances of some class of objects. This is very concise and conveniently exploits the similarities of frames and objects.

In previous work, we have described a set of generally applicable physical sensor features [7]. The manner in which physical sensors convert physical properties to some alternative form, i.e., their transducer performance, can be characterized by: error, accuracy, repeatability, drift, resolution, hysteresis, threshold, and range. These properties can be encoded in the appropriate slots in the frames describing the sensor.

3.2 Sensor Specification

An object-based style of programming requires that the logical sensor be re-described in terms of objects and methods. We shall next give the general flavor of this style, but it must be remembered that any particular sensor is actually an instance of some object class, and, in fact, inherits properties from many levels up.

Thus, in order to get data from a logical sensor, the *characteristic output vector method* must be invoked. Likewise, to issue control commands to the sensor (e.g., camera pan and tilt parameters), the *control commands method* must be used. The role of the *selector* is still the same as in previous logical sensor implementations, however, it now, in essence, is invoked to produce the characteristic output vector.

Such a representation makes it very easy to design sensor systems. Moreover, such specifications allow for replacement of sensors and dynamic reconfiguration by simply having the *selector* send messages to different objects. Given current object-based programming technology, such systems can be rapidly developed and permit dynamic typechecking (on objects).

A logical sensor specification defines a grouping of algorithms, sensors, etc. This newly created logical sensor is an instance of the *logical sensor object* and can be sent messages. As mentioned above, there are two methods defined on logical sensors: the *characteristic output vector method* and the *control commands method*. Thus, any logical sensor can be defined recursively in terms of other logical sensors (including itself).

Currently, our main interest is in the automatic synthesis of logical sensor specifications. Given a CAD model of an object, we would like to synthesize a specific, tailor-made system to inspect, recognize, locate or manipulate the object. Note that the synthesis of a logical sensor specification consists, for the most part, of interconnecting instances of sensors and algorithms to perform the task. This is done by writing the selector to invoke methods on other logical sensors. Given certain constrained problems, most notably the CAD/CAM environment, such a synthesis is possible.

3.3 The Simulation of Multisensor Systems

Effective simulation plays an important role in successful system development. A key requirement is the support for hierarchical specification of the system and the ability to perform stepwise refinement of the system. In addition, it is necessary to be able to efficiently emulate realtime software that will eventually be embedded in the system. Finally, it would be quite useful to be able to embed physical components in the simulator in order to monitor the system's operation.

An object-oriented simulation methodology is well-suited to satisfy these goals. The multisensor system, that is, the system being modeled, consists of a collection of interacting physical processes. Each such process is modeled in the simulator by an object, i.e., an independent process. Interactions among physical processes are modeled through messages exchanged among the corresponding objects.

This general paradigm is currently supported in the SIMON simulator developed by Fujimoto [5, 18]. A toolkit approach is used in which the simulator is viewed as a collection of mechanisms from which application specific simulation environments are constructed. We are currently exploring the simulation of multisensor systems in the SIMON environment. Simulation can be accomplished by substituting simulation libraries for the run time libraries.

A crucial aspect of the simulation is the ability to execute specific algorithms on specific hardware. SIMON permits such a direct execution technique in which application programs are executed directly on a host processor rather than through a software interpreter. Performance information is obtained through the automatic insertion of probes and timing software into the program at compile time. These probes perform whatever runtime analysis is required to accurately estimate execution time of basic blocks of code. A prototype implementation using this technique has been developed modeling the MC68010 and 68020 microprocessors. Initial data indicate that application programs may be emulated one to two orders of magnitude more efficiently over traditional register transfer level simulation, while highly accurate performance estimates can still be obtained.

4. AN EXAMPLE APPLICATION: CAD-BASED 2-D VISION

A simple example which demonstrates some of the power of the Multisensor Knowledge System approach is that of CAD-Based 2-D Vision. The goal is to automate visual inspection, recognition and localization of parts using pattern recognition techniques on features extracted from binary images.

The Multisensor Knowledge System stores knowledge about the algorithms, sensors, processors, etc. This knowledge is used by application specific rules. The systems to be synthesized here require that a model be created for the part to be inspected, and that a robust and (perhaps) independent set of features be chosen along with an appropriate distance metric.

There is an offline training component. The new part is designed using a Computer Aided Geometric Design system. A set of images are rendered by the CAGD system giving a sample of various views of the part in different positions, orientations, and scales. These serve as a training set to the Multisensor Knowledge System.

A set of rules (or productions) performs an analysis of the views of the part to select a subset of the total set of possible features. Features are used if they are robust, independent and reliable. Once these features have been chosen, a new logical sensor object is created whose only function is to recognize the given part based on an analysis of the selected features. The part detector is then linked into a particular application (e.g., an inspection task at a specific workcell) by sending a message to the appropriate camera.

We design with Alpha_1, an experimental solid modeling system developed at the University of Utah. For the past few years the Computer Aided Geometric Design group has been involved in a concerted effort to build this advanced modeler. Alpha_1 incorporates sculptured surfaces and embodies many theoretical and algorithmic advances. It allows in a single system both high-quality computer graphics and freeform surface representation and design. It uses a rational polynomial spline representation of arbitrary degree to represent the basic shapes of the models. The rational B-spline includes all spline polynomial representations for which the denominator is trivial. Nontrivial denominators lead to all conic curves. Alpha_1 uses the Oslo algorithm for computing discrete B-splines. Subdivision, effected by the Oslo algorithm, supports various capabilities including the computation associated with Boolean operations, such as the intersection of two arbitrary surfaces. B-splines are an ideal design tool, they are simple, yet powerful. It is also the case that many common shapes can be represented exactly using rational B-splines. For example, all of the common primitive shapes used in CSG systems fall into this category. Other advantages include good computational and representational properties of the spline approximation: the variation diminishing property, the convex hull property and the local interpolation property. There are techniques for matching a spline-represented boundary curve against raw data. Although the final result may be an approximation, it can be computed to any desired precision (which permits nonuniform sampling).

The synthesized logical sensor object merely sends a message to the segment program for Camera 1 (a Fairchild 3000 CCD camera), then sends a message to each of the features used, then sends a message to the distance function object with the appropriate weights. The system has been implemented in PCLS (the Portable Common Lisp System) using objects and methods. The feature calculations are performed by running C code called from within the instances of the feature objects.

5. SUMMARY AND FUTURE WORK

The Multisensor Knowledge System offers many advantages for the design, construction, and simulation of multisensor systems. We have described many of those. In addition, we are currently working on a CAD-Based 3-D vision system. That is, we are developing a set of rules which will evaluate the 3-D geometry and function of any part designed with the Alpha_1 CAGD system. In this way, weak recognition methods can be avoided and specially tailored logical sensor objects can be synthesized automatically. Another area of current research interest is the simulation of multisensor systems. We believe that our approach can lead to very natural, straightforward, and useful simulations which can include native code running on the target processors. Finally, we are also investigating the organization of knowledge in the Multisensor Knowledge Base. Certain structuring of the data may lead to improved or simplified analysis.

6. ACKNOWLEDGMENTS

This work was supported in part by NSF Grants MCS-8221750, DCR-8506393, and DMC-8502115.

References

[1]   Albus, J.
      Brains, Behavior and Robotics.
      BYTE Books, Peterborough, New Hampshire, 1981.

[2]   Bajcsy, R.
      GRASP:NEWS Quarterly Progress Report.
      Technical Report Vol. 2, No. 1, The University of Pennsylvania, School of Engineering and Applied Science, 1st
          Quarter, 1984.

[3]   Booch, Grady.
      Software Engineering with Ada.
      Benjamin/Cummings Publishing Co., Menlo Park, california, 1983.

[4]   Chiu, S.L., D.J. Morley and J.F.Martin.
      Sensor Data Fusion on a Parallel Processor.
      In *Proceedings of the IEEE Conference on Robotics and Automation*, pages 1629-1633. San Francisco, CA,
          April, 1986.

[5]   Fujimoto, R.M.
      The SIMON Simulation and Development System.
      In *Proceedings of the 1985 Summer Computer Simulation Conference*, pages 123-128. July, 1985.

[6]   Henderson, T., E. Triendl and R. Winter.
      Model-Guided Geometric Registration.
      Technical Report NE-NT-D-50-80, Deutsche Forschungs- und Versuchsanstalt fuer Luft- und Raumfahrt,
          September, 1980.

[7]   Henderson, T.C. and E. Shilcrat.
      Logical Sensor Systems.
      Journal of Robotic Systems 1(2):169-193, 1984.

[8]   Henderson, T.C., E. Shilcrat and C.D. Hansen.
      A Fault Tolerant Sensor Scheme.
      In *Proceedings of the International Conference on Pattern Recognition*, pages 663-665. August, 1984.

[9]   Henderson, T.C., C.D. Hansen, and Bir Bhanu.
      The Specification of Distributed Sensing and Control.
      Journal of Robotic Systems 2(4):387-396, 1985.

[10] Henderson, T.C., Chuck Hansen and Bir Bhanu .
      A Framework for Distributed Sensing and Control.
      In *Proceedings of IJCAI 1985*, pages 1106-1109. Los Angeles, CA, August, 1985.

[11] Henderson, T.C. and Steve Jacobsen.
      The UTAH/MIT Dextrous Hand.
      In *Proceedings of the ADPA Conf. on Intelligent Control Systems.* Ft. Belvoir, Va., March, 1986.

[12]   Henderson, T.C., Chuck Hansen, Ashok Samal, C.C. Ho and Bir Bhanu.
CAGD Based 3-D Visual Recognition.
In *Proceedings of the International Conference on Pattern Recognition*, pages 230-232. Paris, France,
    October, 1986.

[13]   Mitiche, A. and J.K. Aggarwal.
An Overview of Multisensor Systems.
SPIE Optical Computing 2:96-98, 1986.

[14]   Organick, E.I., M. Maloney, D. Klass and G. Lindstrom.
Transparent Interface between Software and hardware Versions of Ada Compilation Units.
Technical Report UTEC-83-030, University of Utah, Salt Lake City, Utah, April, 1983.

[15]   Overton, K.
Range Vision, Force, and tactile Sensory Integration: Issues and an Approach.
In *Proceedings of the IEEE Conference on Robotics and Automation*, pages 1463. San Francisco, California,
    April, 1986.

[16]   Shilcrat, E., P. Panangaden and T.C. Henderson.
Implementing Multi-sensor Systems in a Functional Language.
Technical Report UUCS-84-001, The University of Utah, February, 1984.

[17]   Shilcrat, E.
Logical Sensor Systems.
Master's thesis, University of Utah, June, 1984.

[18]   Swope, S.M. and R.M. Fujimoto.
SIMON II Kernel Reference Manual.
Technical Report UUCS-86-001, University of Utah, May, 1986.

SENSOR FUSION

Volume 931

Addendum

The following papers, which were scheduled to be presented at this conference and published in this proceedings, were cancelled.

[931-07] **Sensor data fusion**
P. S. Pentheroudakis, Lockheed Aeronautical Systems Co.

[931-09] **Sensor fusion and the submarine periscope**
D. Jones, Kollmorgen Corp.

[931-34] **Exploitation tools in a multisensor environment**
B. Fitch, R. Tracy, General Dynamics Electronics

The following paper was presented at this conference, but the manuscript supporting the oral presentation is not available.

[931-10] **Fighter cockpits: automating the office**
E. Adams, McDonnell Douglas Aircraft Co.

AUTHOR INDEX